North Sea innovations and economics

Proceedings of the conference organized by the Institution of Civil Engineeers and held in London on 20 January 1993

Thomas Telford

Conference organized by the Institution of Civil Engineers in conjunction with the Offshore Engineering Society, and sponsored by *Offshore Engineer*

Organizing committee: R. Goodfellow, Mott MacDonald and Chairman of the Offshore Engineering Society (Chairman); H. Williams, Heerema UK Services Ltd; P. Blair-Fish, John Brown Engineers and Constructors; and B. Smith, SLP Engineering)

A CIP catalogue record for this publication is available from the British Library

Classification
Availability: Unrestricted
Content: Collected papers
Status: Established knowledge
User: Offshore engineers, project managers, contractors, structural engineers

ISBN 0 7277 1953 X

First published 1993

© The Authors and the Institution of Civil Engineers

All rights, including translation, reserved. Except for fair copying, no part of this publication may be reproduced, stored in a retrieval system or transmitted in any form or by any means electronic, mechanical, photocopying, recording or otherwise, without the prior written permission of the Publications Manager, Publications Division, Thomas Telford Services Ltd, Thomas Telford House, 1 Heron Quay, London E14 4JD.

Papers or other contributions and the statements made or the opinions expressed therein are published on the understanding that the author of the contribution is solely responsible for the opinions expressed in it and that its publication does not necessarily imply that such statements and/or opinions are or reflect the views or opinions of the organizers or publishers.

Published on behalf of the organizers by Thomas Telford Services Ltd, Thomas Telford House, 1 Heron Quay, London E14 4JD.

Printed in Great Britain by Redwood Books, Trowbridge, Wiltshire

Contents

Chairman's introduction. R. GOODFELLOW	1
Drilling technology for advanced recovery. G. KING	11
Lifting jackets over subsea templates - advances in substructure design. P. M. BLAIR-FISH	21
The efficient use of structural castings for lifted jackets. A. M. WOOD	33
AMETHYST and HYDE: a generation apart. R. E. MILLS	47
A systematic approach to developing FPF hull configurations. R. C. DYER and B. J. CORLETT	61
Gannet's experience in reducing costs. J. H. T. CARTER	77
Economics of pipeline bundles. A. PALMER	85
Elf Enterprise Caledonia Piper/Saltire pipeline bundles. T. W. TACONIS	94
Ensuring the integrity of ageing offshore piplines using on-line inspection tools and fitness-for-purpose methods. J. C. BRAITHWAITE and P. HOPKINS	112
Engineering aspects of jacket toppling as a means of platform abandonment. S. WALKER and J. WILLIAMS	132
Open forum	147

Chairman's introduction

R. GOODFELLOW, Mott MacDonald Ltd, UK

The existing North Sea offshore oil and gas developments are establishing an infrastructure of structures and pipelines in the Northern, Central and Southern regions.

Future field developments will be smaller (less than 50 million barrels of recoverable reserves). Over 100 undeveloped discoveries are within a 30km distance of a production facility. Satellite and marginal oilfield development is restrained by

   (a) high capital expenditure (CAPEX)
   (b) deep water (in excess of 200m)
   (c) reservoir characteristics (satellite wells, gas lift)
   (d) short field life (7-11 years).

The challenge is to find innovative ways to economically exploit these smaller marginal fields (see Figure 1).

Options for development of satellite fields

The main technical challenge is to transport the produced fluids to a production facility for processing and export.

Options for the transport of products are

   (a) direct flow of the multiphase fluid to the host platform
   (b) direct flow of the multiphase fluid aided by downhole pumps
   (c) direct flow of the multiphase fluid plus a subsea slug catcher
   (d) direct flow of the multiphase fluid mixed with stabilised crude oil (Wellstream Absorption Transportation)
   (e) direct transport of multiphase fluid with gas liquid separation
   (f) Direct transport of multiphase fluid boosted by subsea multiphase pumps.

Subsea separation

With subsea separation of the gas and liquid phases the products may be transported along separate lines as follows

   (a) transported unaided

North Sea innovations and economics. Thomas Telford, London, 1993

NORTH SEA INNOVATIONS AND ECONOMICS

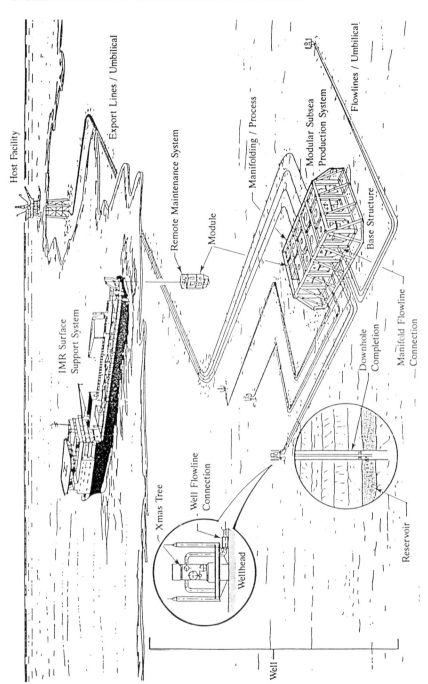

Fig 1. Production and support systems building blocks and sub-systems definition

CHAIRMAN'S INTRODUCTION

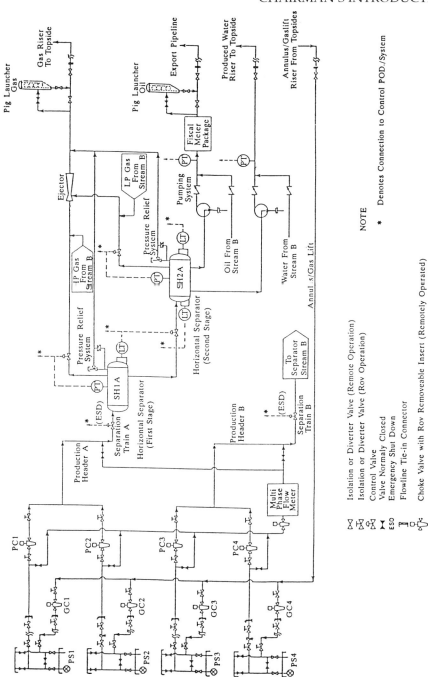

Fig. 2. Simplified flow diagram

(b) gas transported unaided but liquid pumped, using phase pumps
(c) both products boosted using subsea compressor (gas) and single phase booster pumps (liquid).

In addition subsea separation of oil, gas and water may be carried out and the products transported in separate lines either unaided or with subsea booster pumps. A prototype subsea separator was evaluated at the Teeside test-centre. A Simplified Flow Diagram is shown in Figure 2.

Subsea booster pumps

With the capability of boosting the pressure of the produced fluids for transport to the host platform, low flowing wellhead pressures can be tolerated subsea. By comparison, for development with wellheads topside at the platform, head and flowing pressure loss related to water depth limit the products life of the field. The same well produced subsea can have a significantly longer production life as lower flowing wellhead pressures can be tolerated. A typical extended production and improvement in the recoverable reserves as an example of the impact of subsea boosting is shown in Figure 3.

Modularisation

During the field life some of the primary units of a subsea system will need to be retrieved for maintenance.

To minimise the operation expenditure, locating units in retrievable modules is necessary particularly in deepwater diverless systems. A layout of a modularised system is shown in Figure 4.

Modularisation introduces the need for connections for product lines and controls. An innovative solution for this need is the valved multi-port connector (VMC); a full scale prototype was tested as part of the GA-SP projects (see Figure 5). (Acknowledgements to the Bardex Subsea Corporation).

Economics

A profitable rate of return (ROR) (see Figure 6) is paramount for any field development, and with smaller marginal fields the evaluation of all options is essential.

Factors which affect the rate of return and economics are

(a) tax regime
(b) oil/gas price
(c) exploration and drilling costs

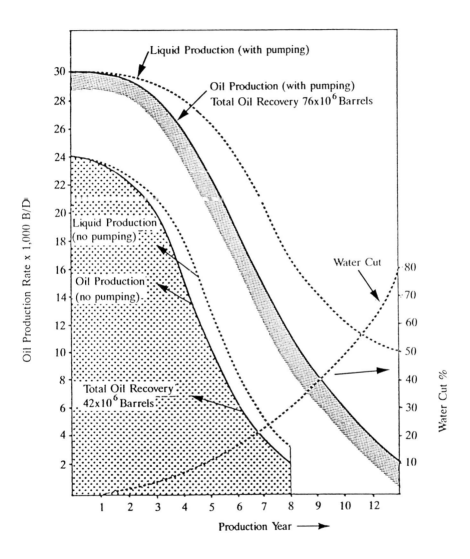

Fig. 3. Production profile: example of impact of subsea boosting on total recovery

NORTH SEA INNOVATIONS AND ECONOMICS

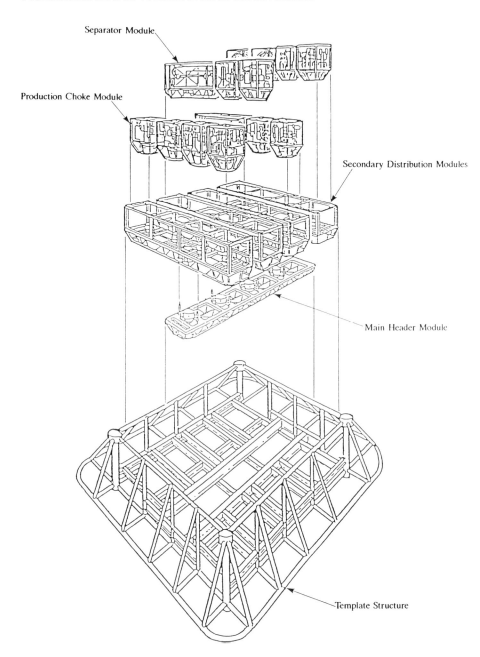

Fig. 4. Modularised system

CHAIRMAN'S INTRODUCTION

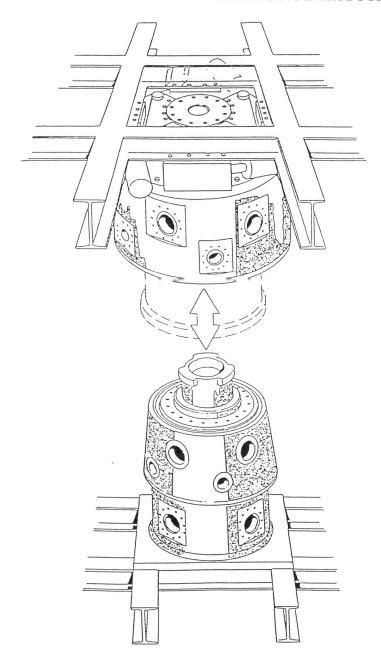

Fig. 5. Valved multiported connector

NORTH SEA INNOVATIONS AND ECONOMICS

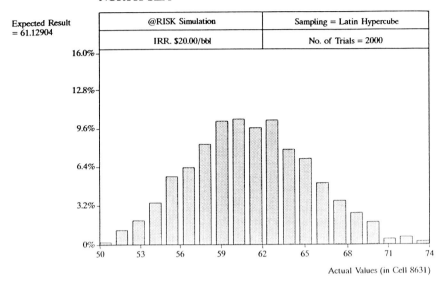

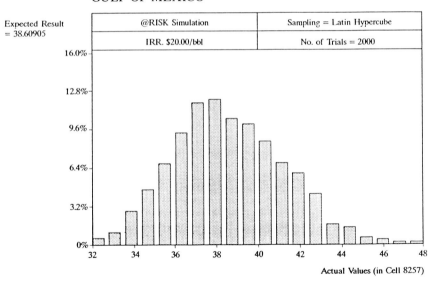

Fig. 6. Rate of return

CHAIRMAN'S INTRODUCTION

NORTH SEA

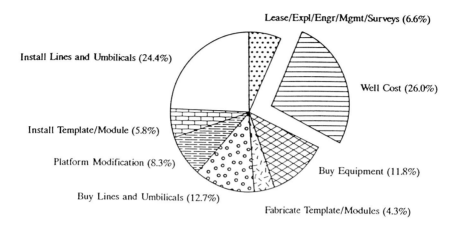

GULF OF MEXICO

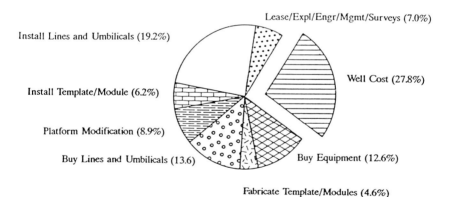

NOTES

1. The areas of these pie charts are set in proportion to capital investment ie. the North Sea chart has a diameter 11.4% greater than that of the Gulf of Mexico.
2. Capital Investment, 1991 Dollars.

Fig. 7. Capital cost estimates

(d) capital expenditure (CAPEX)
(e) production life
(f) operating expenditure (OPEX)
(g) abandonment.

The demand on engineering innovation is to lower the CAPEX and minimise the OPEX. A comparison of capital costs for a subsea development (North Sea versus Gulf of Mexico) is shown in Figure 7.

This conference addresses these problems and offers solutions to ensure the economic viability of future field developments.

Drilling technology for advanced recovery

G. KING, Mobil North Sea Limited, UK

Introduction
This paper is a review of the innovations in drilling technology which have had an economic or enabling impact on the development of North Sea fields. The paper considers the advances in key drilling technologies; for example directional drilling and surveying tools, in the context of the prevailing development pattern of the North Sea. The paper concludes with an overview of the emerging technologies which will have an impact on the next generation of North Sea fields.

When considering the importance of 'drilling' in all its forms to North Sea developments there are two critical points to consider. Drilling is an integral part of all phases of any offshore development, from exploration through appraisal to the drilling of the development wells, and finally to the well abandonments at the end of the field's life. Drilling also accounts for a large percentage of the overall development costs for a North Sea project, and so can determine the profitability of a development. The paper considers the technological innovations and applications which have contributed to the economic development of North Sea oil and gas fields, through the history of North Sea developments and through the development cycle of an offshore reservoir.

There are many sources for the information included in the paper and it would not be practicable to identify them all. Key references are given to papers which document a specific application of technology.

The history of North Sea drilling
In common with most industries, the drilling business strives for continual improvement: wells need to be cheaper, deeper, further away. These requirements have led to a change in drilling technology. In order to have a framework in which to examine the evolution of technology, the drilling history of the North Sea has been divided into approximate areas, corresponding to the major development projects of the time.

NORTH SEA INNOVATIONS AND ECONOMICS

Drilling technology from the mid 1960s to the mid 1970s

The first era of North Sea drilling was dominated by the discovery and development of the gas fields offshore from East Anglia and the oil fields of the Central North Sea and East Shetland basin. The drilling technology utilised for these projects was largely unremarkable. The technology base was imported from the Gulf of Mexico drilling in the USA, and to a lesser extent land drilling within the UK.

Existing technology was usually adequate for the projects; there was a need to bring fields on stream as quickly as possible and, perhaps most importantly of all, there was no indigenous North Sea culture of the application of technology. All the factors led to the wholesale adoption of existing methods, with the result that innovation was minimal. However it is worth placing this in the context of an industry where the basic principle of rotary drilling has remained unchanged for a very long time.

Mobile drilling rigs

There was one area of drilling technology where the requirements of the North Sea did lead to new developments. The prolific East Shetland Basin fields were in deeper water and a harsher environment than the industry had explored before. The small semi-submersible drilling rigs and drill ships from the Gulf of Mexico which were used for the early wells were barely adequate for summertime working, and often incapable of working throughout the year. The requirements of the North Sea drilling industry led to the construction of the 'Second Generation' of semi-submersible drilling rigs, typified by the successful Aker H3 design.

Drilling technology from the mid 1970s to the mid 1980s

Through the latter half of the 1970s and into the 1980s the scale and problems of North Sea development drilling were crucial in the evolution and adoption of new technologies. This was the era when the major North Sea fields were being brought on stream with multiple platforms, often with two drilling rigs on each. The level of drilling activity was sufficient to encourage the development of new technologies which can now be identified with the North Sea.

Drilling fluids

In rotary drilling drilled cuttings are removed, and the open borehole supported by a drilling fluid pumped down the rotating drill string with returns up the annulus. Drilling fluids were originally fresh water viscosified with dispersed cuttings; later the drilled solids were removed with surface equipment and the desired rheological properties obtained by the addition of clays and chemicals.

The geological formations overlying the North Sea reservoirs contain thick claystone sequences, which can react with fresh water to produce a thick 'gumbo' paste which can result in drilling problems until ultimately the drill string becomes stuck in the well. To overcome the problems, 'Oil Based Muds' (OBM) were used more extensively offshore than had been the case before. The early OBMs used diesel as the base oil, which was superseded by mineral oils of lower toxicity. OBM had a dramatic effect in reducing drilling costs through the improved performance of drilling bits and in the reduction of hole problems caused by swelling shales.

OBM drilling fluids were seen as a panacea for all drilling conditions and by the mid 1980s were in widespread use for all well types. The proliferation and concern over the environmental effects of dumping drilled cuttings contaminated with oil has led to the gradual replacement of OBM with a new generation of water based drilling fluids.

Drill bits

Rotary drilling became established with the development of the roller-cone rock bit. Prior to this two bladed bits had been used which quickly became blunt and were unsuitable for hard formations. In the 1970s machine tool cutting technology in the form of 'Polycrystalline Diamond Compacts' (PDC) was adapted to create a new generation of fixed cutter bits. The combination of PDC bits and OBM drilling fluids led to a dramatic improvement in the drilling performance of North Sea wells.

Directional drilling and surveying

A typical platform for an East Shetland Basin oil field may have up to 60 well slots and a drilling radius of 2.5km. Thus two wells starting from surface locations within 2m of one another may reach reservoir targets more than 5km apart. To achieve this requires that the well paths be known and can be controlled, and that the well being drilled does not intersect an existing well.

Directional drilling was an established technology long before the development of the North Sea fields. However the very high cost regime of North Sea operations and the complexity of the directional drilling requirements encouraged the rapid development of new technologies. Two technologies where development was centred on the North Sea were 'Measurement While Drilling' (MWD) directional survey tools and high accuracy well bore survey tools.

Until the mid 1970s directional wells were surveyed by interrupting the drilling operation to run a photo recording magnetic survey tool down the drill string to the bit. The tool measured the inclination and azimuth of the drill string and recorded the results on film. Surveys could take more than an hour and there was always the risk of the drill string becoming stuck in the hole while the pipe was left stationary as the survey was being taken.

Directional surveying was transformed with the development of MWD tools. The MWD tool is a drill collar (a thick wall section of drill pipe run just above the bit to keep the drill string in tension and apply weight to the bit) containing

(a) a turbine alternator assembly linked to a battery to provide power;
(b) an array of magnetometers and inclinometers to measure the inclination and azimuth of the MWD tool;
(c) a poppet valve assembly in the bore of the collar which can temporarily shut off the flow of drilling fluid.

To take an MWD survey, the string is left stationary and the pumps are switched off. This will initialise the tool and take the survey, the pumps are then started. the survey data is transmitted to the surface as a series of pressure pulses up the column of drilling mud, caused by the opening and closing of the poppet valve. The pressure pulses are decoded by a transducer in the circulation system on the surface. MWD tools allow directional surveys to be made with the minimum of interruption to the drilling operation, resulting in more up time and a reduced risk of hole problems.

In order to plan well paths which avoided existing wells it was essential to know the precise location of the wells close to the surface where the separation distances were the smallest. This requirement led to the development of a unique North Sea survey tool which used the inertial navigation systems developed for military uses in aircraft and missiles, with an accuracy far superior to the traditional systems. The tool allowed the close proximity well planning essential for the large North Sea platforms with 40 - 60 directional wells.

Case history

The Forties field was initially developed by BP with four platforms in the early 1970s and brought on stream in 1975. In 1984 the development was extended. A fifth platform was placed over a sub-sea template of 14 wells which had been pre-drilled with a semi-submersible drilling rig.

The original development wells were drilled with 'traditional technology' with an average drilling time of 40 days for the 82 wells. The 14 additional wells used many of the innovations described above: OBM drilling fluids, PDC bits and MWD tools, and had an average drilling time of 29 days (ref. 1).

Drilling technology from the mid 1980s to 1992

The mid 1980s were a turning point in the drilling history of the North Sea. The large scale platform developments had all been completed, leaving 'marginal' fields and outlying reservoirs, and the price of oil fell dramatically, curtailing the expansionist development of technology.

However the drilling industry adapted to the new requirements for drilling more cost-effective wells to poorer quality reservoirs. This is epitomised by the transition of horizontal well drilling technology from a research project to a routine operation.

Horizontal drilling

One of the well parameters which will determine well performance is the length of the reservoir penetrated by the well: the longer the exposed reservoir interval, the greater the inflow of oil or gas can be. Thus it is possible to achieve high production rates with the minimum pressure gradient across the reservoir.

Horizontal drilling technology was pioneered by Elf in France in the early 1980s and demonstrated with the development of the Rospo Mare field on the basis of the improved productivity of horizontal wells. The practice of horizontal drilling is essentially an extrapolation of existing directional drilling technology with the benefit of new rotary drilling technology. There are now many examples of North Sea fields where horizontal wells have been essential to achieve economic production rates (refs 2 - 5).

Rotary drilling technology

Rotary drilling requires the transmission of surface rotary drive to the drill string through some intermediary device. Until the mid 1980s this was achieved by using a flat sided joint of drill pipe, the 'kelly', suspended from a swivel which allowed the passage of drilling fluid. The 'kelly' was rotated by a set of bushings which permit vertical movement of the drill string while transmitting rotary motion from a rotating bushing set in the drill floor. Once the length of the 'kelly' had been drilled, the 'kelly' was pulled to the surface and a new joint of drill pipe was connected to the top of the drill string. This meant that drilling was continually being interrupted, and it was not possible to rotate or circulate through the drill pipe while it was being run in to the hole.

High angle and horizontal wells require the ability to continually circulate mud through the drill pipe to keep the hole clean and to rotate the drill pipe to prevent it from becoming stuck. The continual interruption of the drilling process and the inherent limitations of rotary drilling with a 'kelly' led to the development of 'Top Drive Systems' (TDS).

In a TDS the drill pipe is attached directly to a large powered swivel which is suspended in the derrick. This allows the driller to drill using the full height of the derrick and so drill in 30m intervals rather than the conventional 10m lengths for 'kelly' drilling, as well as allowing circulation and rotation at any time. The TDS technology has been critical for the drilling of horizontal wells and has also led to significant performance improvements for conventional wells (ref. 6).

A refinement and adaptation of existing drilling technology has been the introduction of 'steerable drilling systems' (SDS). A typical SDS comprises a positive displacement motor which converts the hydraulic power of the drilling fluid into rotary motion of the drill bit, and an angular misalignment section. The misalignment is configured to give the drilling assembly different directional drilling performance, dependent on whether the drill string is being rotated or held stationary. By varying the amount of pipe rotation it is possible to drill a precise well path with small course corrections without the need to change the entire drilling assembly. There are also surface adjustable drilling assemblies which may be changed to give different directional drilling performance.

Slant hole drilling

In the 1980s the Morecambe gas field in the Irish Sea was developed using slant hole drilling rigs as a result of the combination of a shallow reservoir and a shallow water depth. The conventional 'J' profile of a standard deviated well with a vertical origin was inappropriate for the Morecambe field: the drilling radius of each platform would have been too small at 600m for an economic development.

The drilling radius could be increased by 75% if the wells were to be given a surface inclination of 30 deg. Thus the field was developed using inclined drilling conductors and slant drilling rigs (ref. 7).

Extended reach wells

The drainage area of a platform increases with the second power of the drilling radius: doubling the drilling radius will quadruple the drainage area. Thus there is a significant benefit in drilling wells to a greater distance from the platform. The holder of the current record for the greatest radius is a Statfjord 'C' well with a departure greater than 6km and a total length greater than 7.2km. This well combined almost all the preceding technological developments, and demonstrates how significant improvements can be achieved by combining technologies rather than developing new solutions (ref. 8).

Deep high pressure wells
In parallel with the development of drilling technology for wells at a greater distance, the technology was being developed for drilling wells to greater depths. The central North Sea has become a highly prospective area for deep high-pressure (surface well pressures of 12,000 psi during well tests) gas condensate wells. The drilling and testing of such wells in the harsh environmental conditions of the North Sea now make the region a world centre for the technology.

Drilling safely and cost-effectively to great depths with high formation pressures has required the development of a number of key 'enabling' technologies. The most important of these have been in the development of well bore casing and well head systems capable of withstanding 15,000 psi internal pressures (refs 9 - 11).

Drilling technology for the future

Drilling technology for North Sea needs has evolved with a combination of innovation and optimisation. The steady improvement of existing practices will continue, and will inevitably result in more cost-effective wells. However well costs are constrained by a legacy of huge past investments in traditional technology and traditional drilling practices. In order to make a significant reduction in the well costs, in the order of 50%, some new and radical technology will be required. Without being fanciful, it is possible for us to identify two technologies which are already at the prototype phase of development and which have the potential to make a significant reduction in drilling costs.

Slim hole drilling
The generic North Sea combination of hole and casing sizes (36"x30"; 26"x20"; 17.1/2"x13.3/8"; 12.1/4"x9.5/8"; 8.1/2"x7") is both difficult to explain and hard to justify. Drilling costs could be reduced dramatically if smaller holes could be drilled and there was a smaller difference in diameter between hole sections. For example 8.1/2"x7"; 6"x5"; 4"x2.7/8" hole and casing combinations could be used. Such wells have been drilled to 2000 - 3000m on land. The challenge will be to take the technology both offshore and to greater depths.

Multi-lateral drilling
It is possible to produce two vertically separated reservoirs with the same well by isolating the two zones and running two strings of production tubing. The technology to produce two reservoir sections which are separated hori-

zontally is the next challenge. The objective is to drill multiple wells out of a common well bore, and then produce the wells simultaneously. The technology is a development of the drain hole drilling which has been done in the past (ref. 12).

Conclusions
The drilling technology for advanced recovery currently used in the North Sea has been developed to provide cost-effective solutions to drilling challenges. The pattern of the application of technology is one of lag and led: where drilling technology has been inadequate, so a new solution has been developed; the new technology can be applied in other areas, making new developments economic. It is reasonable to assume that this pattern will continue and technology will continue to develop.

Acknowledgements
The author would like to thank the management of the Mobil North Sea Limited Drilling Department in Aberdeen for their support in the preparation and presentation of the paper.

References
1. DENHOLM J.M. North Sea Development Drilling South East Forties: A Case Study. Paper SPE 15448 presented at the 61st Annual Technical Conference of the Society of Petroleum Engineers, New Orleans, October 5 - 8 1986.
2. CLARK A.C. and COCKING D.A. The Planning and Drilling of the World's First Horizontal Well From a Semi-Submersible Drilling Rig. Paper SPE 18712 presented at the 1989 SPE/IADC Drilling Conference, New Orleans, February 28 - March 3 1989.
3. KOONSMAN T.L. and PURPICH A.J. Ness Horizontal Well Case study. Paper SPE 23096 presented at the Offshore Europe conference, Aberdeen, September 3 - 6 1991.
4. PEDEN J. Horizontal Wells - Their application and Economic Benefit. Paper presented at Conference for Developing Cost Cutting Strategies for Offshore Production, Aberdeen, 20 - 21 May 1992.
5. ANDERSEN S.A. et al. Horizontal Drilling and Completion: Denmark. Paper SPE 18349 presented at the SPE European Petroleum Conference, London, October 16 - 19 1988.
6. BOYADJIEFF G.I. An Overview of Top-Drive Drilling Systems Applications and Experiences. Paper IADC/SPE 14716 presented at the 1986

IADC/SPE Drilling Conference, Dallas TX, February 10 12 1986.
7. WHITE S.M. and BARTLE M.F. Slant Drilling in the Morecambe Field, Irish Sea, U.K. Paper SPE/IADC 16151 presented at the 1987 SPE/IADC Drilling Conference, New Orleans, March 15 - 18 1987.
8. NJAERHEIM A. and TJOETTA H. New World Record in Extended-Reach Drilling From Platform Statfjord C. Paper IADC/SPE 23849, presented at the 1992 IADC/SPE Drilling Conference, New Orleans LA, February 18 - 21 1992.
9. LOW E. and SEYMOUR K.P. The Drilling and Testing of High Pressure Gas Condensate Wells in the North Sea. Paper IADC/SPE 17224 presented at the 1988 IADC/SPE Drilling Conference, Dallas TX, February 28 - March 2 1988.
10. KRUS H. and PRIEUR J.M. High Pressure Well Design. Paper SPE 20900 presented at Europe 90, The Hague, 22 - 24 October 1990.
11. SEYMOUR K.P. The Next Step - Planning a 20,000 psi Well for the UKCS. Monograph presented at the SPE HPHT seminar, Aberdeen 14 May 1992.
12. CROWDEN M. Texaco's Teal Prospect Produces a Record Well. Offshore/Oilman, July 1992.

Discussion

The speaker was asked whether slim hole drilling technology would require the design and fabrication of new drilling rigs, or could this new drilling technology be used on existing drilling rigs. In his response the speaker indicated that he could not see anyone investing money in the building of new drilling rigs and that it was more likely that existing drilling rigs would be used. In addition he posed the question as to whether new drilling rigs were required at all, as it may be possible to drill from platforms. He added that he felt that retrofit rigs should be used to perform slim hole drilling or appraise areas where well costs would be high.

A delegate enquired whether it would be (a) possible and (b) desirable to drill through different structures now that the capability to drill over long distances has been achieved. In his response the speaker said that it is indeed possible to produce from two vertically separated reservoirs.

Another delegate wanted to know whether the adoption of the slim hole drilling technique would in fact reduce the range of extended drilling. The speaker agreed that the ability to do both would not be possible, as enough casing strings could not be fitted to provide enough weight. He suggested the idea of zero stiffness since one would be considering a drill bit approx 4" in diameter at the end of a 7km casing.

The speaker was then asked whether risk analyses played a part in the economics of the decision to drill further, and the possibility was suggested of encountering a problem approximately 1km from the final location of the drilling operation. The speaker responded by saying that risk analyses formed a part of the overall strategy.

The speaker was asked how Horizontal Drilling and Extended Reach Drilling compared with respect to time and costs. The speaker stated that if it was going to take 120 days to do the drilling of a 6-7km extended reach well, then one would have to consider the cost of this compared with not having vertical access to the well drilled from a template, with perhaps 40 days of drilling time. However as different fields have different requirements then one may have to commit a lot of money up front or consider using both types of drilling.

A delegate referred to an article on re-use of drilling rigs and package rigs in *Offshore Engineer* and was interested in any comments on this. The speaker quoted the example of the drill rig from Claymore being moved to Beatrice, and added that as we are now moving to an environment where facilities will be re-used the drilling package should be made more modular and as re-usable as possible.

Lifting jackets over subsea templates - advances in substructure design

P. M. BLAIR-FISH, John Brown Engineers & Constructors, UK

Introduction
There were two major innovations in the design of steel substructures for North Sea platforms in the 1980s. The first innovation was to place a subsea template one or two years before the platform was installed, so that wells could be pre-drilled by a jack-up or semi-submersible drilling rig. Over a dozen such templates have been engineered and installed. The second major innovation is to install steel jacket substructures by lifting, rather than launching, in water depths of up to 175m. Lifted jackets use the capacity of the largest semi-submersible crane vessels. Lifted jackets were developed to reduce the costs in response to the fall of the price of oil in 1986. Less temporary steel is used in temporary buoyancy and launch frames. To date, over a dozen large lifted jackets have been engineered and fabricated. This paper describes how such structures are installed, and compares them with conventional barge-launched structures. Cost and weight savings are discussed for vertical piles, swaged connections, expanded plan bay spacings, and innovative materials for mudmats and coatings.

The economics of offshore oil and gas are significantly improved if costs are reduced or project timescales shortened.

Two major innovations in the design of fixed steel substructures have significantly improved the economics of offshore developments in the North Sea in the 1980s.

Firstly, subsea templates have been installed on the seabed one or two years before the platform is placed over them, so that wells could be pre-drilled by a jack-up or semi-submersible drilling rig. Secondly, the increased capacity of semi-submersible crane vessels has enabled the substructures of up to 10,000 tonnes to be installed by lifting rather than launching.

Subsea templates under fixed platforms
A typical template under a fixed platform has (see Fig. 1)

(a) receptacles for wellheads

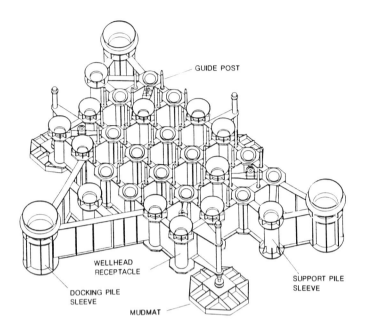

Fig. 1. Large drilling template

(b) guide posts for lines to guide the lowering of a BOP stack
(c) mudmats to support the template on the seabed
(d) levelling system
(e) support piles to secure the template during drilling
(f) docking piles to position the substructure over the template.

Support piles and mudmats must not obstruct wells to be drilled from the platform, unless the mudmats are removed before the substructure is installed. Docking piles must line up with sleeves attached to the substructure. Also, the substructure must not foul on the template.

Early templates supported wellhead receptacles indirectly between their main framing. On recent templates structural framing runs directly into the wellhead receptacles as shown in Fig. 1. This change was made to reduce fabrication costs. Wells are generally spaced at 2.59m (8' 6") centres. Lesser spacings have been used, despite tight clearances between BOP stacks and capped wells.

The height of the template depends on the length of the wellhead receptacle, and on the need to keep well caps clear of drill cuttings and grout. The overall weight of a template is typically 12 tonnes per slot, excluding piles. The weight may increase if docking piles, mudmats or pin piles are awkwardly placed so as to avoid a clash with platform wells. Conversely, the weights may reduce if a proprietary unitised template such as Cooper Tools or McEvoy is used.

Installation of templates

The template is installed by lowering to the seabed on the drill string of a jack-up or semi-submersible drilling vessel, or on the underwater block of a crane vessel (see Ref. 1). Jack-up rigs can only handle small lightweight templates. Semi-submersible drilling vessels can install compact templates of up to 16 wells as described by Sedco-Forex (Ref. 2). Multi-service support or heavy lift vessels are needed for larger templates.

Templates can be levelled by screw jacks operated from a drill string or by hydraulic power supplied by an umbilical to the surface. Levelling jacks either react against mudmats or grip and climb up support piles.

Support piles may be installed by drilling and grouting, or driving, depending on the equipment available. However, drilling and grouting is slow for the larger (1.5 - 1.8m OD) piles usually required for docking piles. If a crane vessel is used, piles can be installed by driving, either using above-water hammers and followers or using underwater hammers. Also, a crane vessel can handle piles as single pieces.

After driving, support piles are connected to the template either by grouting or mechanical swaging. Grouting needs longer sleeves and the assistance of either divers or a nimble ROV. Mechanical swaging has been developed by BUE Hydralok through prototype tests on 1067mm OD x 25mm WT piles in a Scottish Loch in 1984 to fatigue tests on model piles (see Ref. 3).

Mating of substructure with template

To avoid lateral loads on wellheads during mating, docking piles are usually disconnected from the template by removing their guide sleeves.

Small jackets have been mated to templates by an open box in the jacket framing that fits over and around a guide cage on the template. One large jacket has been mated horizontally with bumpers and latches, but vertical mating is now more common because the installation is much quicker. The male/female system on the jacket may be a fixed sleeve, an active system where pins are lowered (Ref. 4) or a passive system. A passive system with sliding sleeves and cones on top of docking piles (see Fig. 2) can be designed to minimise excitation of jacket motions by installation waves close to a natural frequency of the jacket as it is ballasted and lowered into its final position.

Motions and forces during mating can be predicted using time domain analyses (see Ref. 5) that have been bench marked against the results of model

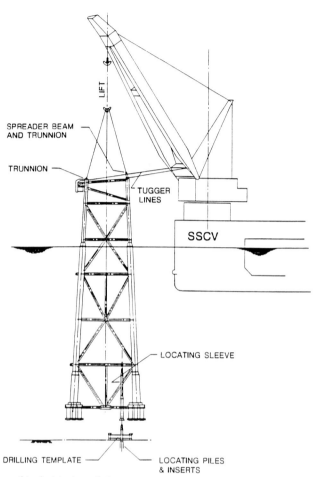

Fig. 2. Mating of jacket to template

tests. Motions at the base of a jacket are generally small, but may increase if the jacket is hook upended. Sideways forces during mating are generally limited to 2MN (200 tonnes) to avoid overstressing docking piles. As a result, allowable sea states for mating are generally of the order of Hs = 2m for barge launched jackets and Hs = 1-1.5m for lift-installed jackets.

Drilling and tieback of template wells

Template wells are generally drilled through a 762mm (30") conductor set beneath the template, a Blow Out Preventer supported on the first inner casing and a marine riser up to the drill rig. Wells are capped after drilling. After the structure is installed the wells are tied back to the topside, generally

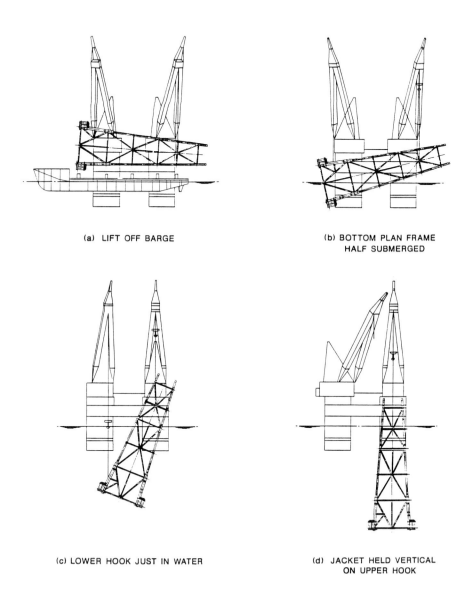

Fig. 3. Installation sequence for BP Unity Riser jacket

by a 508mm (20") conductor. Tie-back conductors are supported laterally by oversize guides which allow tie-back tools to pass through. The alignment of the tie back conductor over the well generally needs to be within 300mm in plan and 1.5 degrees of verticality. Jackets can generally be set to within 0.5 degrees of true vertical.

Economics of pre-drill templates
A typical 12-slot template costs of the order of £1½ million for materials and fabrication and of the order of £2½ million to install. Wells drilled through a template from a semi-submersible or jack-up rig will cost more than wells drilled from the installed platform. The cost of the template and the extra cost of pre-drilled wells has to be set against the revenue from earlier production. Tie-back of pre-drilled wells enables production to start shortly after the platform is installed and hooked up. Investment in two drill rigs rather than one per platform also speeds drilling but is considerably more expensive.

Lift-installed jackets

Barge-launched jackets have launch frames and temporary buoyancy that are required for temporary conditions. Some designs minimise temporary buoyancy by self-upending (Ref. 6). Some conceptual designs even seek to launch jackets from their main legs. However, re-use of temporary buoyancy is not often cost-effective. Therefore, weight and cost can be saved if launch frames and temporary buoyancy are avoided altogether.

The semi-submersible crane vessels DB102 and Micoperi 7000 were designed to lift topside modules but can lift and install slender jackets weighing up to 10,000 tonnes (see Refs 7, 8 and 9). For water depths of up to 70m, jackets may be built and lifted upright, as for the 8768 tonne BP Gyda jacket which was installed in 66m of water in 1989. For water depths of 75-175m, jackets have been installed by tandem lifting followed by hook assisted upending. Over a dozen such jackets are now installed in the North Sea.

Installation sequence
Up to 120m water depth, jackets can be lifted off a transport barge (see Fig. 4) and upended by lowering the base of the structure until the spreader bar for the rigging to the upper hook passes over the tops of the jacket legs, as for the BP Unity Riser Platform (see Fig. 3). Where possible, the lower hook is released before it enters the water. For water deeper than 120 metres a two-stage lift is required with some re-rigging after the initial lift into the water. Auxiliary buoyancy was required for the Veslefrikk jacket in 175m of

water (see Ref. 7). The final lowering of the jacket onto the seabed on the hook of one crane of a semi-submersible crane vessel helps location over a pre-installed template (see Fig. 2). The safe sea state for lift is typically Hs = 2m. Motions and forces during lift off the transport barge can be predicted by analytical methods which have been bench-marked against the results of model tests.

Configuration of lift-installed jacket

Lift capacities and clearances are greatest if the plan dimensions of the jacket are minimised. A lift-installed jacket will show greatest savings compared to a launched jacket if both top and bottom plan dimensions are smaller than those of a launched jacket. To date only two tandem lift-installed jackets have any base dimension greater than 45m.

Load-out of a lifted jacket is usually by skidding on skid shoes or a purpose built frame. To minimise pitch and heel of the cargo barge and crane vessel during lift the centre line of the jacket is usually on the centre line of the cargo barge. Also, the centre of gravity of the jacket is usually in line with the centre line of the crane vessel for the initial lift.

Lift points for the initial lift off the barge need to be connected to the main framing of the jacket. Lift capacity is greatest if there is an even distribution of load between the cranes. If possible, the upper lift points should be located so that the spreader bar on the upper rigging can pass over the tops of the jacket legs (see Fig. 3).

Plan levels are therefore set by lift geometry as well as the need for acceptable intersection angles between braces and legs, and to support conductors, risers, facility caissons and J-tubes. Acceptable spacings between upper bays are generally governed by vortex shedding and fatigue. Conductor spans in lower bays, however, are governed by stability or vortex shedding. Guidance on design for fatigue and vortex shedding is summarised in Ref. 10. Stability of conductors can be assessed assuming that loads due to internal casings do not contribute to buckling. As stated by Stahl and Baur in OTC 3902 (Ref. 11), the masses of internal strings do not change their potential energy due to deflections of the conductor. Following load tests on model conductors, the interaction equation proposed by Stahl and Baur was rewritten in terms of allowable loads:

$$U = (P_e/P_{allow}) + (P_i/(0.6 \times P_y)) + 0.85(M_e + M_i)/(1 - P_e/P_{allow})0.66 M_y$$

where P_e is external axial load at point of interest due to BOP stack, etc., P_i is internal tensile load, P_{allow} is $P_{cr} \times 12/23$, P_{cr} is critical axial buckling load, P_y is $F_y \times$ (area of section), M_y is $F_y \times$ (elastic section modulus), M_i is moment

Fig. 4. Lift of Shell Gannet 'A' jacket

due to eccentricity of internal casings, M_e is moment due to wind and wave loads, platform movements and eccentricity of BOP stack.

Weight and cost savings

The weight and cost of lift-installed jackets have been reduced by the use of vertical piles, leg inserts to resist boat impact and novel systems for corrosion protection and mudmat support.

Vertical piles driven by underwater hammers had already been adopted for barge-launched jackets (Ref. 6) to save the weight and cost of pile guides. However, anodes and grouting systems on pile sleeves must be designed to resist vibrations during pile driving.

Soft surface soils require large mudmats. Mudmats at such sites can be made lighter by the use of aluminium and by pile grippers (Ref. 7). However, cost savings from aluminium mudmats are as yet unproven. Alternatively, the jacket may be landed on pre-installed piles (Ref. 12). However, the cost of the marine operations for the pre-installed piles may outweigh the savings from the smaller mudmats.

The legs of a jacket should be strong enough to resist an accidental impact from a supply boat, particularly for a lift-installed jacket with only four legs. Thick-walled tubulars are generally cheapest, but offshore-installed grouted inserts (see Ref. 7) reduce lift weight and give greater overall strength.

Corrosion protection by bare steel and anodes may account for up to 5% of the installation weight of a jacket, and increase wave drag by the order of 10%. Weight and wave loads can be reduced by painting the entire structure so that the anodes become smaller. Wave loads may be further reduced by the use of anti-fouling coatings on the upper part of the jacket (Ref. 7).

Weight can be saved by using high strength (grade 450) steel in areas governed by strength rather than stability or fatigue. Fabrication costs may be reduced by a rational approach to the need for Post-Weld Heat Treatment, based on analyses and test of fracture toughness.

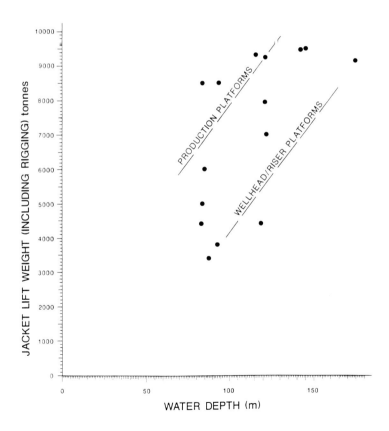

Fig. 5. Jacket lift weight versus water depth

Economics of lift-installed jackets
The lift weights of lift-installed jackets depend on water depth, topside load, and the presented area of appurtenances at waterline. The variation of lift weights with water depth is shown in Fig. 5. The typical lift weight of 9,300 tonnes for a lift-installed jacket for a production platform in a water depth of 125m is around 25% less than for a comparable barge-launched jacket. If slings and spreader beams are re-used and the same crane vessel is used to install a lifted as a launched jacket, then the lifted jacket is of the order of 20% cheaper than the launched jacket.

Conclusions
Pre-installation of subsea drilling templates under fixed structures has enabled early drilling of platform wells. Lift-installed jackets have reduced costs. Both innovations have contributed to the economics of projects.

Acknowledgements
Permission from BP and SHELL to publish details of the Unity Riser and Gannet 'A' jackets is gratefully acknowledged. Also, the author thanks his colleagues at John Brown Engineers and Constructors for constructive comments.

References
1. SKILTON L.C. Sub-sea drilling templates. *Steel Construction Today*, 1992, Vol. 6, No. 2, March, 68-72.
2. HAMPTON J. E. Improvements in or relating to the transportation of subsea templates. UK Patent GB 2068 439B.
3. LOWES J. M., HOLMES. R and PEEL J. W. Performance of a swage pile/sleeve connection when subjected to cyclic loading. *Offshore Technology Conference* paper OTC 6950, 1992, 45-53.
4. LENIHAN C., AUSTIN R.T.C. and FLANAGAN P.J. The rapid installation of a large North Sea jacket over a subsea template. *Offshore Technology Conference* paper OTC 4759, 1984, 419-430.
5. DUNCAN P.E., DRAKE K.R. and WINDLE D. Jacket to template docking forces: predicting full scale loads from results of physical and computational models. *Behaviour of Offshore Structures*, Trondheim 1988, 891-906.
6. FRANKLIN J.L. and GUNN N.J. Eider Jacket Design Sets Cost Effective Example. Ocean Industry, 1986, February, 39-41.
7. BLAIR-FISH P. M., BAERHEIM M. and AUSTIN C. W. Design of the Veslefrikk wellhead jacket. *Steel Construction Today* 1989, Vol. 3, No. 4,

August, 103-108.
8. MATHARU J.S. Design of the Gannet A jacket. *Steel Construction Today*, 1992, Vol. 6, No. 2, March, 51-54.
9. CROWLE A.P. Heavy lift, from concept to installation. *Installation of Major Offshore Structures and Equipment Conference*, February 1993.
10. BARTROP N.D.P. and ADAMS A.J. *Dynamics of Fixed Marine Structures*. Marine Technology Directorate, 3rd Edition, 1991.
11. STAHL B. and BAUR M.P. Design Methodology for Offshore Platform Conductors. *Offshore Technology Conference* paper OTC 3902, 1980, 465-478.
12. PATTEN R.B. and VENTER K.V. Pre-installed piles for large offshore structures. *Transactions of the Institute of Marine Engineers*, Vol. 103, pp 63-72.

Discussion

A speaker referred to a slide which showed the potential savings between lifted and launched jackets and claimed that the actual savings were in fact much higher than shown. In his response the speaker said that the comparison was run cautiously in order to show that a lifted jacket would cost less than a launched jacket and that one would really expect a saving in the region of 20-25%.

A delegate posed the question of whether one could expect to see an increase in the weight of crane vessels, thus giving increased lifting capacity. Responding, the speaker said that such a vessel would allow one to lift jackets in deep water conditions and could thus be of benefit in the frontier areas. He added that a 30 000 tonne vessel could help in the area of decommissioning.

A delegate commented that he felt the frontier areas would be too exposed for heavy lifts to be practical. The speaker acknowledged that launched jackets were less sensitive to sea state conditions than lifted jackets.

The speaker was asked how the cost of fabricating and installing two liftable jackets on the Bruce project compare with that of one jacket. The speaker responded by saying that the concept of using two jackets had already been conceived at the stage John Brown became involved, and conceded that while it may have been possible to combine the two jackets, the combination of the two topside layouts would have proved more difficult.

A delegate then asked that, as lifted jackets were now at the limits of the geometry of the crane vessels, how did the speaker see the future for these vessels? The speaker said that these vessels would have to follow the market and this would mean taking the lifted jacket concept to other parts of the

NORTH SEA INNOVATIONS AND ECONOMICS

world. He added that if no new vessels are built then one could probably lift jackets in water depths up to 200m in the North Sea.

Another delegate asked about the use of mating guides rather than pins on templates. The speaker stated that general practice was to develop a system whose lateral stiffness varied as the jacket was lowered onto the template, so that resonance between the sway period of the pinned jacket and waves was avoided.

The efficient use of structural castings for lifted jackets

A. M. WOOD, River Don Castings Limited, UK

Introduction

Over the past 10 years, castings have been used to an ever increasing extent in offshore structural applications, although quite naturally this has been limited to specialist areas of use. The benefits of using castings are the stress-reducing and fatigue advantages which stem from the presence of a smoother simplified joint, and of course the economic advantages which go hand in hand with this. Initially their use was mainly in topsides with the need for simplified heavy lift padears and complex column base and stab-in nodes.

A major influence on the increasing use of castings for steel jackets has been the recent popularity of crane installed structures. For the heavier jackets at least, cast padeyes
trunnions and spreader bar ends have been desirable on both economic and technical grounds. The tilt installation method has also encouraged the use of cast hubs rotating on machined trunnion spigots.

Besides these lift points, the substitution of castings for fabrications has also been driven by the need to solve fatigue problems, usually at face brace X nodes, conductor framing or leg K brace nodes near the splash zone. Castings have also helped solve problems with boat impact loads, and to a lesser degree but still important has been a desire to combat spiralling fabrication costs.

We have been involved in the design and supply of cast lift points for all the major lifted structures from Arbroath, Kittiwake, Veslefrikk and Gyda some years ago, to the more recent Gannet, Nelson, CATS, Bruce, Scott, Sleipner and Dunbar projects. The total of 148 nodes and 56 spreader bar ends supplied by River Don Castings represents over 90% of all the castings used on jackets.

Undoubtedly, the opportunity to reduce the frequency of in- service node inspection and the potential reduction in long term repair problems for welded nodes are also considered to be important by some designers/oil companies.

Part of the success of castings has been due to innovative engineering ideas and a dedication to achieving short schedule times and delivery reliability.

Fig. 1. Padeye shear plate for heavy lift jacket

The following discussion contains only limited reference to the metallurgy, weldability and mechanical properties of cast node steel, as this is considered by now to be a well proven subject. We have delivered nearly 15,000 tonnes of this vacuum treated, fine grained, C/Mn - 1% Ni steel, and its good weldability and low temperature toughness properties are well known. Characteristics of this steel are the through thickness properties in sections up to 600mm. A well proven high yield strength version is now available ($400 + N/mm^2$) which is further discussed below, and the weldability and -50 degree C impact properties are excellent.

Innovative uses of castings

Padeye shear plates

We first proposed this through leg cast shear plate concept (figure 1) for Kittiwake and it marked the move away from 'stuck on' fabricated padeyes, and thus simplified the lift point node. This design has now been up-rated for the heaviest jackets and optimised to minimise weight. The best version incorporates a T or KT stub group integrally with the casting on the underside of the leg, together with a portion of the leg chord and parts of any ring stiffeners which may be necessary.

Trunnion nodes

Trunnions which are offset from the jacket legs or the X brace face nodes and, together with a rotating hub, allow the jacket to swing through 90 degrees into the install position, have also necessitated cast solutions. The early types used a cast insert into the leg chord, but this was not the most ideal situation because high stresses still occurred in the interface region of the fabrication, and complex and expensive stiffening was still required. More recent designs have included the full leg chord circumference as a casting and also integral ring stiffeners. Alternatively, a complete face X node, with trunnion can be made as a casting (figure 2), and low weight versions (approximately 40T) are now available for the heaviest lifts. Fabricated node equivalents would certainly weigh over 60T. The plain cast chord option without stubs is really only advantageous if very tight schedules apply, or if stub sizes cannot be confirmed early.

Fig. 2. Complex face X node incorporating trunnion

Fig. 3. Deck support Y node for jacket

Spreader bar ends

These are now invariably made as castings, since they would be extremely complex and expensive to make as fabrications. Again, these were first used for Kittiwake and the optimised compact design is enhanced by necking where possible. We now have a number of standard designs and patterns available 'off the shelf' for capacities up to about the lift limit of existing cranes.

Y-junctions and support nodes

Jacket designs with complex framing and more than 4 legs, may include tubular Y-joints. Where the member angles are acute, there will be technical advantages in substituting a cast joint, either as a whole node, or even just as an insert, to cover the most inaccessible area for welding. The BP Gyda jacket used large Y node castings and also complex Y-junction support nodes at the

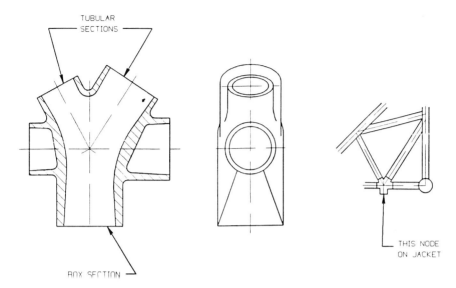

Fig. 4. Cast Y node used for jacket support/skidding

top of the structure to efficiently transfer the high topside loads into the structure (figure 3). The same concept can also be used in reverse, as support for a jacket during construction and load-out. The lower section changes to box section and the casting efficiently transmits forces to the launch rails (figure 4). A high load bearing cast Y-joint can also be used in a structure with twin stubs uppermost to accept stab-ins from two closely positioned decks or modules (figure 5).

Leg K brace nodes

Two projects in Norway (Statoil Veslefrikk and Europipe) are benefiting from the use of cast K-joints to combat fatigue in the splash zone regions of the jackets.

X Nodes - simple and complex

The first castings to be used on jackets, nearly 10 years ago, were small X nodes for three shallow water gas installations, where their role was to improve fatigue life. Since those early days, X nodes have been used on a further 7 large lifted jackets. Besides their fatigue benefits, castings also have advantages in resisting high boat impact forces. Special chord reinforcement can be adopted to minimise the potential yielding when a full plastic moment is applied at the node stub ends. It would be difficult to design a welded node to survive such loading.

Besides the simple X nodes described above, 20 or so large complex face brace X nodes (figure 6) have been used over the past three years, particularly on the BP Miller and Bruce projects. Besides their main benefit in solving fatigue or boat impact problems, in some cases these nodes have also been convenient places to site lifting trunnions or buoyancy tank mounting pintles. Open access throughout the centres of the nodes into the stubs has also proved a benefit during fabrication by reducing the extent of radiography necessary for closure welds.

Novel X node

River Don Castings has recently patented an unusual form of X-node which has at its centre a solid ring (figure 7). Besides the potential to make considerable weight and cost savings over conventional fabricated nodes (see below), it also has a cost saving advantage associated with manufacture.

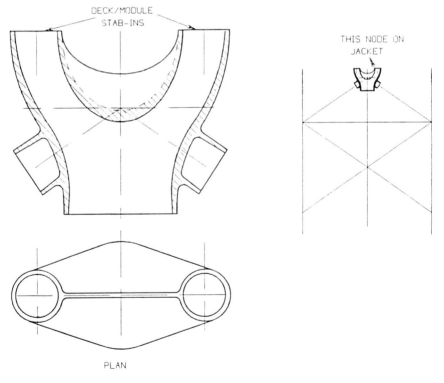

Fig. 5. Cast Y node used for deck/module support

Because the stubs of the node have a common work point at the centre of the ring and the pattern equipment has separate stubs which can slide around the ring core, then only a limited amount of pattern equipment is necessary to cover a wide range of X node geometries. Different stub diameters can also easily be incorporated, and thus the number of expensive patterns can be minimised. The solid ring core assists in feeding and is sized to ensure adequate bending resistance. This design concept has been proven by FE analysis and there is a good chance that it will be adopted on jackets in the near future.

The design optimisation process for castings

The process of optimisation always starts with a review of the geometry of incoming members and manipulation of work point eccentricities to allow, wherever possible, the maximum shrinking of the joint size. In some cases this shrinking process can give dramatic weight savings (see below). Whilst engineering efficiency is a basic aim, the castability and ease of 'feeding' of the solidifying casting, to eliminate shrinkage defects, is also a critical consideration. Simplification is also an important feature because for practical reasons manufacturing must be within a sensible timescale, and schedules are becoming ever tighter.

Fig. 6. Complex face brace X node for jacket

Necking of the incoming members, or the centre of the joint, sometimes even to the point of having a solid central core rather like a pin joint, can be an option for the designer. This concept is one of a number of options which we always consider wherever appropriate. In fact this feature is covered by a patent which is held by River Don Castings. Often, this necking process can be the key to minimising weight, but it is by no means a universal panacea and there are many joint geometries where it is not at all appropriate. Unless a jacket is of very regular design with limited geometry variations, it is unlikely that one concept, such as the pin joint, will be universally appropriate. In some cases, the ring node concept described below will be better.

The approach to design optimisation is a complex one because it is influenced by many factors, including a number of practical foundry ones which will not be common knowledge. Also, we can rely on detailed design evidence from hundreds of previous examples, mostly including FEA stress results, so this has a strong bearing on design decisions.

Design basis for fatigue

As with fabricated nodes, the most common approach for castings has been to achieve a target SCF so that a given design life can be reached based on a particular design line. The welded T-line has been the most commonly used design line to date because it is known to be conservative with respect to the performance of cast material. Thus in such cases, the fatigue life benefit for a casting comes only from the SCF reductions due to smooth radii and evenly tapering sections towards stub ends. This beneficial effect alone is quite significant, since it is clear from our FE analyses of dozens of nodes that peak SCFs (invariably at the weld prep detail) are usually below 2 and often below 1.5. Thus four-fold fatigue life improvement over welded joints is easily achievable. Specific design lines for castings have been proposed previously by ourselves and others showing around 10- fold improvements in design lives over fabrications in the working stress range region. More recently independent verification of this design line has taken place, and the Health and Safety Executive will shortly publish a recommended design line showing fatigue benefits of 4 to 30 for castings, depending on stress range level. It will of course not be possible to use anywhere near this full fatigue advantage, by designing thinner section nodes, because of practical foundry limitations.

Weight and cost savings due to casting substitution

The novel X node concept already described is a good example of how, for certain fabricated nodes with large bottle chords, the substitution of a casting can considerably save weight. In one situation recently studied the weight was reduced from 19 tonnes to 12 tonnes, including stub length corrections (figure 7).

Even with fabricated node prices as low as £3,500-£4,000/tonne, which is unusually low for such complex nodes, the casting offers financial savings. In addition of course there are the fatigue and reduced in-service inspection benefits which will accrue. However, it should be stated that such large weight savings are only possible for complex, bottle chord nodes. For simple X nodes it may only be possible to achieve weight and cost equivalence.

Potential weight savings from the use of high yield strength cast steels

River Don Castings have now developed a 420 MPa yield strength cast steel, and this is now available as a well proven material (CSN 400+). It has already been supplied to a number of customers for castings with major section thicknesses. The development work has included cut-up trials covering sections up to 500mm thick, and mechanical property results are shown in Fig. 8 and Table 1. Previous high yield casting experience has always encountered problems in ensuring high yield strength, good weldability and good low temperature impact and fracture toughness resistance in thick sections. However, such problems have now been solved, and we have produced castings weighing up to 25 tonnes in this grade, and larger ones are planned.

Unfortunately, the potential for saving weight with high yield steel can sometimes be limited, because practical foundry thickness rules override structural thickness requirements. Also, for example with trunnions, the need for certain sling sizes trunnion radii, etc., will be dominant factors in sizing the casting. Padeye plates are, however, a good example where high yield steel can give a clear and significant weight saving benefit. Whilst for certain nodes, it is possible to achieve 10-15% weight saving, it is unlikely that the 30% theoretical saving could ever be reached for large nodes.

However, the option for a high yield strength casting is worthwhile, because the weld prep interface detail is improved when meeting a high strength member.

Conductor guides

For a number of years River Don Castings, along with certain jacket designers, have proposed the possibility of cast conductor guides to reduce fatigue cracking problems and possibly also reduce costs.

Two basic design types are available. The substitution approach replaces the now common cantilever type of fabrication, but the use of a casting allows

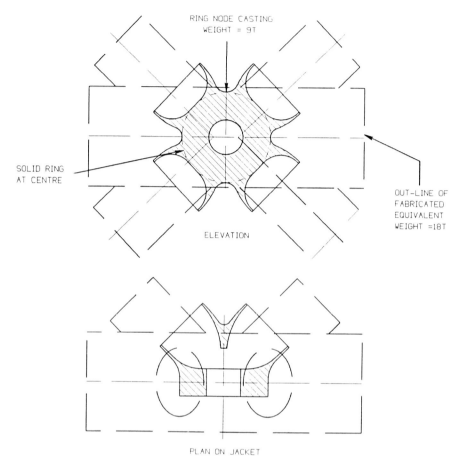

Fig. 7. Cast novel X ring node: size and weight reductions for jacket face node

Table 1. CSN 400+ CTOD valves at -10 deg C, 100 mm^2 specimens

PARENT METAL mm	WELD METAL mm	HAZ mm
0.265	1.420	0.986
1.093	1.483	1.045
0.748	1.445	0.797

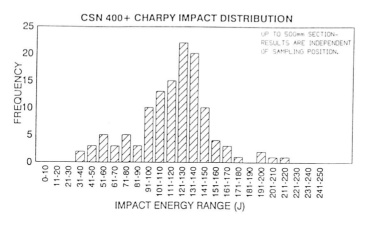
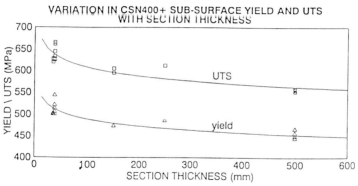
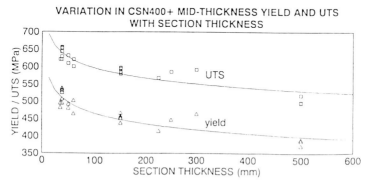

Fig. 8. CSN 400+

the guide to tuck in very close to the support member. Thus where fatigue problems exist due to the size of support members, their size can be maximised without creating welding problems. The connection detail to the parent tubular can also be made so as to maximise fatigue life.

Alternatively, a cast conductor guide can be centred on a brace member, or the intersection of two members. It would then be similar in whole or in part to the Novel X node concept already described, and the consequent direct load paths could be an advantage to the jacket designer.

The future for castings

The cost minimisation process for steel jackets is now in full swing, with a clear understanding amongst all parties that novel low-cost solutions are needed if the pace of North Sea development is to be maintained. We are addressing this already by concentrating on the following

(a) Pursuing ever more efficient light-weight designs, and pioneering the supply of high yield strength castings.
(b) Minimising manufacturing overheads with a large throughput of castings made to tight schedules.
(c) Considering other new casting design options such as Novel X and Cast Conductor Guides to extend our product range.

We are also gearing up for further increases in supply capacity, as well as responding to the faster build times which are essential for a low cost steel jacket. In this respect, we have a particular cost and speed advantage in that many of the numerous existing designs/patterns are appropriate for re-use. For example we have

(a) 15 standard padeye plate designs
(b) 12 trunnions
(c) 12 simple X nodes
(d) 20 spreader bar end types.

Another way to reduce jacket costs, would be the adoption of a greater degree of design standardisation, and one operator is now pursuing this in the form of a standard large jack-up rig as a platform. On the same theme, we think there are possibilities for standard jacket concepts which could apply to deeper waters and for higher topside loads. Some years ago, a hybrid concrete/steel self-floating structure was proposed which eliminated leg batter angle and combined the benefits of both types of structure. Figure 10 depicts such a hybrid, made as a two part structure, with only the upper section being lift- installed by a relatively small capacity crane vessel, and

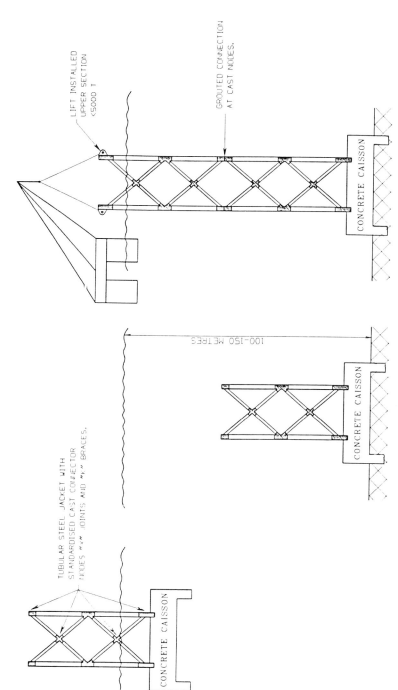

Fig. 9. *Concept for standardized medium water depth jacket with repeatable cast nodes (minimized lift installation capacity and elimination of piling)*

also with the need for piling eliminated. Such a structure would utilise standard cast X and K nodes for fatigue resistance and also cast leg connector nodes at the interface with the concrete and at the stab-in junctions between the two steel sections. Such a simplified well head or drilling structure could also be coupled to a floating production system moored alongside.

Acknowledgements

The author wishes to thank the Directors of River Don Castings Limited for permission to publish this paper and also colleagues at River Don Castings for their assistance.

Discussion

A speaker enquired whether stubs would be added on in the cast x-ring node in comparison with the more conventional alternative. In addition he also asked would River Don Castings promote cast conductor guides for all levels, as fatigue was only a problem close to waterline. In response the speaker said that there would indeed be stubs to add on. In relation to the conductor guides besides fatigue benefits there could also be cost savings over fabrications. He said that the conductor design would be governed by loads from a dropped conductor as this is the worst case, and all levels should be considered for castings.

A delegate wanted to know whether there were savings to be obtained in terms of the repeatability of the use of the moulds. The speaker said that the pattern was reusable, perhaps 40-50 times, but not the mould. A repeatable design from the same pattern could give savings of 10-15% in the cost of the casting.

Another delegate commented that in the past there used to be a lot of cutting out and welded repairs on the castings, and asked how the present castings fared in this respect. The speaker responded to this query by saying that numbers of defects were generally shape-dependent. Whilst progress had been made in minimising weld repairs, they were still necessary in many cases, but were included in the cost and schedule of the delivered casting.

A delegate enquired as to the interaction between the designer, operator and fabricator such as River Don Castings. The speaker said that often the foundry gets involved at a late stage of the design/fabrication because many standard shapes are now available. Ideally however they would prefer to be consulted at the earliest possible stage.

AMETHYST and HYDE: a generation apart

R. E. MILLS, Kvaerner H&G Offshore Limited, UK

Introduction

There has never been more pressure to develop oil/gas projects for still lower costs. The decline in value of oil and oil- related products in real terms, coupled with major difficulties in reliably forecasting development and operating economics over the life-of-field, is forcing operators to seek significantly reduced Capex and Opex costs in order to assure viable economic return. This situation is neither peculiar to the development of offshore projects nor the oil-gas industry since much of the underlying pressure for change associated with international and domestic trading conditions is common to most sectors of UK industry.

Other constraints peculiar to the UK offshore industry have been the decline in new field size available within UKCS, increased focus on small gas-field developments, enhanced recovery of hydrocarbons in-place made more difficult by inhibiting geological characteristics, and particularly the overhaul in safety standards brought about following the Cullen Report.

Not-Normally-Manned (NNM) platforms have received increasing attention from UKCS operators in recent years. This interest comes from the fact that NNM platforms often provide improved performance, and a more added-value safety-compliant, quality-sufficient and cost-effective basis than conventional options for the development of smaller fields or field extensions that are an increasingly significant part of UKCS activity. However, without recent developments in associated instrumentation and communications technology, it is doubtful that the NNM platforms concept would be considered sufficiently reliable.

Neither is this interest only for gas field developments. Elsewhere NNM platforms have been adopted for oil and gas/condensate production and such a progression is a possibility for the North Sea. Indeed the extension of exploration into increasingly hostile and remote areas is likely to add to the attractiveness of NNM installations.

The NNM approach brought significant advantages to the AMETHYST project in the application of new technology and optimisation of conceptual opportunities available. These aspects are reviewed below. However, of more particular interest is the radical upshift in the application of both human and company engineering which is currently unfolding on the HYDE project. Successful realisation of the fundamental elements inherent in this novel strategy within the HYDE project can open new possibilities for other developments; the opportunities are not confined to NNM platforms. Indications for the future both in terms of NNM platforms approach and more especially in the projection of HYDE-experiences are tremendously attractive and exciting. Clear vision and strong commitment will be necessary on the part of corporate organisations and personnel involved if such possibilities are to be appreciated and fully realised.

NNM platform classification

Formal UK Department of Energy classification of NNM platforms began only with the British Gas Exploration and Production Limited Morecambe Field, Stage 2 development.

By definition, the NNM concept implies that normal day-to-day platform functions, such as routine production control, must be carried out automatically and remotely without the need for operator presence on the platform. Less-usual functions, such as local ESD checks, planned shutdowns, major maintenance, statutory checks and surveys, require personnel visits to the facility. The timing and frequency of visits to NNM classified platforms is generally defined and limited by the criteria given in Table 1.

The correct balance must be struck between the number of personnel

Table 1. Criteria for personnel visits to NNM platforms

- No personnel sleep or live on the platform.
- Maintenance crews aboard during daylight hours only.
- Overnight stays limited to emergency conditions only.
- Normally no personnel aboard the platform.
- Defined, non-routine visits by operational and maintenance personnel allowed.
- Helicopter must be on-call when personnel are aboard NNM platform.
- Helicopter pre-flight procedures strictly enforced to anticipate severe weather conditions.
- DoEn regulations restrict visits to fewer than three days in any week.
- Maximum personnel per visit determined by (lower of) helicopter or lifeboat capacity.
- Standby vessel must be within 5mm.
- Routine maintenance must be minimised (including simplification of instrumentation and electrical distribution).

visiting an NNM platform, the frequency of visits and platform safety features.

While the main design features of NNM platforms emphasise reliability and the ability to sustain enhanced reliability under remote control, the normal absence of operating personnel drastically reduces the risks associated with long term exposure of people offshore.

AMETHYST

The AMETHYST field was acquired with many other North Sea oil and gas interests as a result of the take-over by BP of Britoil in 1988. AMETHYST was originally discovered in 1972 but until 1983 it was not believed that it could be economically exploited. A combination of gas price rises and the development in recovery and remote control techniques changed all that.

Britoil did not believe that a manned platform carrying out processing offshore was the most economical development concept - moreover they did not have a land based terminal in the area.

British Gas were commercially selected to receive and process AMETHYST gas at their Rough Terminal at Easington. Britoil were to remain the operator with overall responsibility, but would delegate routine operating, planning/logistics, processing and maintenance to British Gas. Thus BP inherited a situation where Britoil were planning to use British Gas to process gas partly owned by British Gas (one of the partners) in the first place, and wholly destined for sale to British Gas - an unusual situation but not unique in the world of oil and gas production.

Te £260 million AMETHYST development is a relatively small gas field requiring four well sites for full-field depletion. It comprises two trunkline wellhead platforms (which commenced production October 1990) and two satellite wellhead platforms (which commenced production October 1991).

Operational requirements

The economics of a conventional approach with a manned host platform were unattractive, and a remotely controlled multiplatform development was conceived as commercially necessary and optional in terms of safety. The whole success of the chosen development configuration depended upon the simplification of facilities such that no platform would need visiting more than once a week and then only in daylight hours. This was the key operational requirement, as it determined platform maintenance and operations direct costs and limited the need for costly standby boat facilities.

In order to actively drive operating costs down to a minimum the following design criteria were adopted

(a) simplicity and low equipment costs
(b) minimal diversity consistent with flexibilities inherent in sales contract
(c) high risk approach to asset protection
(d) maximising of local infrastructure
(e) reliable, easily maintained equipment
(f) application of proven technology to improve safety and operability
(g) gas processing in an established onshore terminal having spare capacity (opportunity for cost efficiency).

The success of AMETHYST was realised by the incorporation and packaging of a number of new features, which not only produced significant value-for-money development savings, but also resulted in lower operating costs than for conventional fields. The annual operating expenditure for AMETHYST field is less than £7 million.

Key features/issues
Early discussions with the gas purchaser, regulatory authorities and the operations group resulted in the following major issues/key features being adopted within the design basis.

Compression required some 5 years into field life and to be installed onshore (platforms to have no continuous
rotating machinery.
Pipelines (trunk and in-field) operated 'wet' supported by active corrosion control regime.
Remote control and monitoring (local control not allowable or installed).
Proprietary package for the digital distributed control system (with telecomms and control-point diversity).
Simple custody transfer metering offshore (fiscal onshore).
Minimal maintenance power supply (submarine cable from onshore terminal).
No fire-water pumps.
No reliance on HVAC (open-ventilated decks).
Basic shelter facilities (inflatable beds for emergency stop-overs).
Single standby and guard vessel for four AMETHYST platforms.
Guard vessel deployed in early years to ward off errant vessels in busy shipping area.
Low-corrosion steel (13% Cr) well completions.
Corrosion inhibitor/hydrate suppressant mix exported by piggy-back pipeline from the terminal.
Topsides designed for wellhead shut-in pressure (obviates flares - satellite platforms only).

Battery back-up for emergency supplies extended to allow continued production for up to four days.
Hydraulic or electric valve actuators (no plant or instrument air).
Electro-hydraulic ram-luffing crane.
Intelligent digital microwave link (more reliable and faster than analogue).
Production shut-down from onshore terminal upon extended communication loss to any platform.
Simplified topsides facilities (reduces CAPEX, OPEX, hook-up and commissioning, installed footprint, weight, programme and fabrication).
Single-train processing (equipment reliability and control integrity satisfying demand- supply requirements).
Platform safety consideration (since the field is located astride busy shipping lanes).
Three-phase pipeline flow (enabled by low water-cut).
Bespoke substructures designs for each platform (due to differing water depths and sub-soil characteristics.
Higher intrinsic safety due to low inventory of hydrocarbons arising from simplicity of facilities.
Impact on safety facilities and operational procedures owing to remoteness from land or existing platforms.
Portable equipment for wire-line and well servicing.
Integration of control and shut-down systems (reduces maintenance).
Unit replacement rather than lengthy on-platform repairs.
Stock main spares onshore (sparing on field-wide basis).

Advances in technology and equipment reliability, particularly relating to instrumentation and communication equipment, made a significant impact in the development and adoption of NNM platforms on the UKCS. This success derived principally from two factors: improved reliability of a range of equipment and systems and successful implementation of newer technology.

Proven off-the-shelf systems provided additional capabilities for remote control, data acquisition, reporting and fault detection. Stand-alone programmable logic controllers integrated into a distributed control system permitted operational changes. The greater automation afforded by such digital systems results in increased productivity through higher availability and additional benefits in overall efficiency and safety.

The ability to network using an intelligent digital microwave link provided more flexible, reliable and faster data transmission than the corresponding analogue systems. Platform operation is from a master station onshore, using digital communications intelligent network equipment. The system is based on line-of sight electronic transmission between platforms

NORTH SEA INNOVATIONS AND ECONOMICS

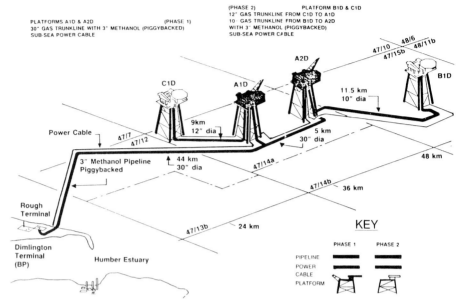

Fig. 1. Offshore installation

whereby one controller can activate/vary any function on any one of the platforms from onshore - even the daily sphere launching to sweep liquids build-up in the pipeline.

With AMETHYST production almost entirely gas, methanol and corrosion inhibitor are injected continuously into the gas stream to suppress corrosion or hydrate formation within the pipeline, which is thus operated 'wet'. A mixture of methanol and corrosion inhibitor is exported offshore along a piggyback pipeline atop the gas export trunkline. Methanol is recovered from the gas and liquids at the terminal for recycling.

Operating the pipeline 'wet' incurs the disadvantage of more frequent sphering or pigging to sweep along liquid build-up. This leads to complexities in launcher design and capacity, entailing more frequent visits to reload the launcher magazine. Daily sphere launching is needed for the AMETHYST trunkline.

Other major but less obvious considerations contributing to low capital field-development costs resulting from smaller topsides and jackets include

(a) wider choice of fabrication yards, often leading to keener pricing or scheduling

(b) topsides fabrication under cover giving less risk of weather interruptions

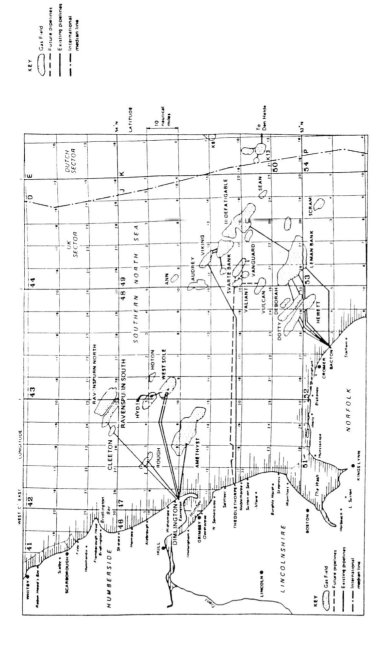

Fig. 2. Field location

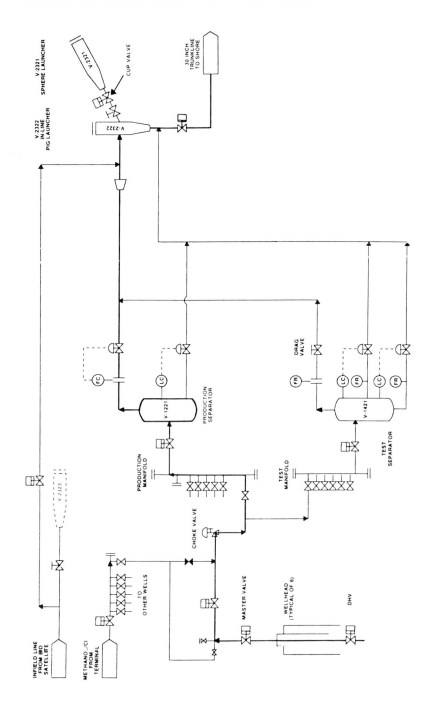

Fig. 3. Typical platform process flow diagram

(c) wider choice of installation and hook-up contractors. Better scope definition giving greater opportunity for lump-sum sub-contracts.

Operational feedback

The production profile over the first 20 months indicates a shortfall of 1% in gas supply in the first contract year, falling to only 0.6% in the second year.

Shutdown investigations have confirmed previous external NDT work, and they indicated that the facilities were in excellent condition after functioning at higher than expected pressures and rates.

Monitoring of installed corrosion probes whilst monitoring modifying C.I. injection rate has shown no hydrate problems and low aggregate corrosion levels.

Two summer programmes of reservoir surveillance were completed successfully within the NNM criteria - each well activity being completed within daylight hours.

Achieved visit frequencies have exceeded the target figure of once per week due to familiarisation activities and teething problems (maloperation of sphere launcher, retrofit of certified fire and blast protection around pipeline ESD valves, energy shortfall from hydraulic panels, freezing of level bridles in cold February 1991 weather).

Safety considerations

Normally, safety of personnel is considered in relation to exposure of each individual to risk whilst at work. With a NNM development such as AMETHYST, the total risk is the dominant consideration. This is well illustrated by the consideration of firepump benefits. There is no doubt that firepumps marginally enhance the safety of personnel once safely on the platform. However, they require a level of support and maintenance such that these benefits are outweighed by the extra visit effort for a not normally manned installation.

The AMETHYST facilities are now such that discounting ship collision the greatest risk to personnel comes from helicopter operations, which generate 40% - 60% of the risk. For this reason the pressure is on to keep the visit team small (and therefore multi-disciplined) and the planned visit frequency low - say once per fortnight, perhaps extending even to monthly.

HYDE

Facilities

The HYDE field lies to the north-west of the WEST SOLE field which was the first development in the Southern North Sea. A single platform is required

NORTH SEA INNOVATIONS AND ECONOMICS

Table 2. Comparison of AMETHYST C1D and HYDE facilities

	AMETHYST C1D	HYDE
Jacket Type/Weight	Through Leg Piles	Skirt Piles
Jacket Weight	960 te	1000 te
Topside structure	Basically Identical	
Main Process Vessels	Identical	
Hydrate Suppressant	HP Piped Methanol	LP Bunkered Glycol
Control System	Fisher Dual Highway	
Power	Cable	On-board Small Diesels
Piping Isometrics	80% Identical	

	AMETHYST C1D	HYDE
Pressure, Temperature, Level Instruments (Transmitters/Switches)	44	30
Total Number of Valves - Excluding Packages/Trees	133	141
Number of Actuated Valves	25	27
Number of Pumps/Fans/Prime-movers	6	8
Number of Fire/Smoke/Gas Detectors	55	66
Number of Light Fittings	110	114

to develop the new field and one of the AMETHYST platforms has been used as the design model.

The platform facilities are compared in Table 2. The main process vessels and topsides structural design are similar, and the development strategy has recognised the potential benefits inherent in commonality.

Included in Table 2 is a limited schedule of some key items that determine topsides complexity and maintenance effort. The differences between AMETHYST C1D and HYDE have been occasioned by

(a) on-board power generation and diesel fuel bunkering
(b) hydrate suppressant (glycol) stored/pumped on platform
(c) gap action level control (not modulating valves)

(d) change to I.R. beam-type fire & gas detection
(e) chromatograph installation
(f) produced water dump facilities
(g) post-Cullen safety features
(h) high flow horizontal wells
(i) water depth, sea-bed current and sub-soil characteristics differences
(j) offshore regulations and guidance notes updates.

Overall these differences are relatively small, as shown in Table 2.

Glycol and diesel tank capacities are such that bunkering intervals are in excess of 14 days. Thus the maintenance visit frequency should be substantially reduced from that achieved for the AMETHYST platforms. Some room exists for further reducing visit frequencies by examining the maintenance routines.

The HYDE facilities lend themselves to this approach. For example, the generators are packaged for maintenance-by-removal and pumps essential to production are spared.

Included within the scope of the project are a pipeline and upgrade works to convert Easington Terminal to wet gas operation.

Contract strategy

HYDE differs from AMETHYST more in the area of contract strategy than hardware. An alliance has been formed to realise extraordinary cost performance targets for HYDE well below those achieved for AMETHYST. Since the basic functionality is the same, savings can only come from doing business differently.

The key to the strategy is alignment of the contractor's goals with those of the operator thereby liberating and harnessing their various strengths. The HYDE contract strategy has three main themes.

Shared risks and rewards. A key feature of the alliance strategy is the risk-reward undertaking that will govern division of the cost savings between the parties, provided the project under-runs the mutually pre-agreed budget. Conversely, if the project exceeds budget, the parties will share the additional costs. Only the project out-turn cost will be used to apportion each party's share of the saving or over-run; each company's individual performance will not in itself be the deciding factor. These terms were jointly formulated based on each alliance party's ability to influence capital expenditure. The parties charge costs on an open book marginal basis - with deferred profit and fixed overhead recovery.

Single fully integrated project team. As the contract strategy evolved so did the concept of a single integrated team. Personnel have been seconded from the various organisations. The team includes an unusually small BP presence (but does include a full time operations representative). All functions are undertaken by the one team - including accountancy. The arrangement seeks to avoid, among other things, duplicated effort and delays in facing, communicating and resolving problems, thereby averting costly project changes. Information which hitherto may have been considered confidential is shared. The benefits of this approach are essentially

(a) reduced management overhead
(b) maximum exploitation of design/fabrication/ procurement synergies
(c) co-operation rather than conflict
(d) enhanced confidence between parties
(e) shared objectives
(f) enhanced return
(g) better product.

Minimum conditions of satisfaction. A brief contract regulates the relationship between the parties. The key to this is a high level list of conditions that constitute a successful completion of the alliance effort. Six single-sentence conditions are considered sufficient, e.g. 'Acceptance of the facilities by the operator together with any funds necessary to complete outstanding work'.

The chosen strategy has enabled the following to be achieved.

(a) A team-wide commitment to finding ways of reducing costs, with a target of achieving a 20% saving on AMETHYST norms.
(b) Developed quality control, e.g. the cost of steel QC effort £0.5/tonne; achieved weld defect rate of 0.3%.
(c) Design tailored for ease of fabrication.
(d) Short programme - start of detailed design to first contract gas 18 months.
(e) Trading across traditional operator-contractor boundaries to achieve overall cost savings.
(f) Enrolment of selected vendors/sub-contractors in shared objectives.
(g) Potential profit margins well in excess of normal.

The project team report to a multi-company panel. The panel comprises a single senior-level representative from each alliance party, and functions by consensus.

Challenges for the future

NNM platforms have been proven as valid practical options for gas production on the UKCS. The factors and considerations which have driven development of the NNM platform concept so far may be expected to continue and should become more important throughout the 1990s and beyond. The AMETHYST and HYDE platforms are typical of simple NNM gas platforms. There are significant attractions in safety and in both capital and operating costs in achieving low visit frequency. The following criteria seem to face the mature offshore industry in UKCS in extending low cost developments into other areas.

(a) Use of cost-benefit analysis in the earliest generation of the process scheme.
(b) Single train/unspared facilities as the starting point.
(c) Relaxed production availability criteria.
(d) Adoption of suppliers of equipment/services and quality systems and document management.
(e) Avoidance of customisation, expensive expediting and extensive quality monitoring/assurance
(f) Creative generation of commercial arrangements built upon shared objectives together with a strong downward drive on costs.

Four other factors may influence operators selection of NNM platforms in the future

(a) wellhead fluid characteristics
(b) extension of NNM concept to oil developments
(c) feedback of further reduced OPEX costs with current NNM installations
(d) reduced exposure of operating personnel to harsh conditions in possibly even more challenging circumstances.

Many of the gas-condensate future developments are characterised by higher wellhead pressures and temperatures (15,000 psi and 300°F). Higher wellhead pressures will permit the economic transmission of fluids over greater distances to shore for treatment and recovery without offshore compression. However, an economic balance will still need to be struck between pipeline pressure drop and entry pressure to the transport network.

Acknowledgements
The author thanks BP Exploration, the AMETHYST Field Development Unit Partners, the HYDE Field Unit Partners and the HYDE Project Alliance Parties for permission to publish this paper.
 The AMETHYST Field Development Unit partners were: BP Exploration

(field operator), British Gas Exploration and Production Limited, Enterprise Oil plc, Amoco (UK) Exploration Co., Amerada Hess Limited, Ocean Exploration Co. Limited, Murphy Petroleum Limited, Arco British Limited, Fina Petroleum Development.

The HYDE Field Development Unit partners are BP Exploration (Field Operator) and Statoil (UK) Limited.

The HYDE Project Alliance parties comprise the HYDE Field Development Unit partners, as above, together with UiE (Scotland) Ltd, Kvaerner H&G Offshore Ltd, and JP Kenny and Partners Ltd.

Kvaerner H&G Offshore Ltd (KHG), together with its subsidiary Kvaerner Earl and Wright (KEW), has played a leading role in the development into operation of several NNM installations on the UKCS; past projects include: British Gas Exploration and Production Ltd's MORECAMBE Field (Phase 1, Stage 2); BP (Britoil) AMETHYST Field (Phases 1 and 2); and Ranger Oil's ANGLIA Field.

Current KHG projects include: British Gas Exploration and Production Ltd's NORTH MORECAMBE Project, Shell (UK's GALLEON Field, and BP's HYDE Field.

Discussion

A delegate enquired about the CAPEX comparison on the original configuration of manned platforms. In his response the speaker said that the original configuration was through the Cleeton field using a manned platform with a development cost of £77m. Changes in geological recognition enabled it to be tied back through West Sole. Development costs then dropped to £65m. He argued that as a robust project it produced better rates of return for all.

Another delegate wanted the speaker to explain where the 'Alliance' started and finished, how much BP do up front and how a low CAPEX was achieved by the choice of design and materials. The speaker explained that they were involved with the development from a very early stage. In order to be assured of the viability of the project BP needed to be shown that a low CAPEX was achievable. As the contractor's profit is affected by the economic viability then there is a commitment by all the parties to achieving a low operating cost. The speaker also said that a lot of attention was being devoted to managing the interface between themselves and BP.

One of the delegates wanted to know at what stage the 'Alliance' tied up with the nominated subcontractors. The speaker explained that the alignment within the 'Alliance' was extended to other key parties. He also said that it would be possible to improve on this area should there be another opportunity but that this did not detract from the successful control of subcontractors on this project.

A systematic approach to developing FPF hull configurations

R. C. DYER, BCD Offshore, and B. J. CORLETT, Hurness Corlett & Partners, UK

Introduction

The selection of hull configuration for a floating production facility (FPF) appears to offer a great deal of choice, with various sizes of semi-submersibles, flat-top barges, tankers or novel designs, with or without storage. Poor initial selection of the hull can have a relatively large affect on the feasibility of the project, as the cost of the hull and its sub-systems are a large component of the total cost. The risk on this cost is also high because the design process is complex, requiring a weight and space balance together with an optimisation of the design for mooring system, utilities configuration, storage requirements, etc.

An optimal FPF design requires the complete integration of subsea, marine and process technology, and a failure to recognise that the interaction of these disciplines is different from that on platform projects will cause problems. The global cost of a development is defined during the conceptual and feasibility stages of the project. Hence, at the early stages of the concept definition, an integrated 'total system approach' should be taken.

Although an FPF development contains approximately the same number of basic elements as a solution using a platform, the elements in a FPF scheme are technically and commercially less rigidly connected than those in a platform scheme. Conventional data-based programs can give reasonable answers where there is a historical and logical progression in the designs, but they have not proved ideal when looking at FPF schemes. It is also difficult to rearrange historical data for projects where there is a large degree of inter-relationship between systems which cross boundaries of functional areas.

With experience of successful FPF projects it is now possible to define a more applicable primary subdivision by careful specification of the interface areas. The second step is then to define each area by a set of parameters which can be manipulated to 'size' the area. From these two re-defined stages it is possible to specify more accurately an FPF hull at an early stage.

The North Sea area has particular design problems because of the harsh environment and relatively shallow water (with respect to wave height)

mooring conditions. This has resulted in little cross-fertilisation of technology with other worldwide FPF developments, and a governmental regulatory approach based on platform projects. This has led to an overly conservative approach on some hull design values (ref.l), and lack of a clear approach to selection of semi-submersible mooring systems (ref.2).

Hull selection - general

For feasibility studies it is essential to ensure that the selection process is a quick, simple and iterative process. To achieve this, we have identified the main parameters which affect the selection and have concentrated on their inter-relationships. Some combinations of parameters either preclude certain options (e.g. a semi-submersible cannot be selected if storage is required) or mandate certain options (e.g. a requirement for well work-over or well servicing excludes a weather vaning solution). We have used such constraints to make a rational simplification of the first stage selection process.

It is possible that there may be field requirements which either lead to the use of non-standard solutions, or which require a cost sensitivity analysis for changes in various components. These can be identified and a more detailed procedure can be developed which allows the designer to select different hull sizes within specified limits. Concrete construction also falls into this category, and ref.3 provides a useful starting point.

Introduction - monohull with and without storage
The basic monohull designs are based on the use of either a standard (marine industry) flat-topped barge or a typical tanker cross-section. The selection of a fully mature design results in a design optimised for weight, structural arrangement and construction, through competitive shipbuilding practice. Any alternative or novel scheme will always have to compete against such base case schemes.

The steel weight estimate is made for the flush hull, and includes only the shell, bulkheads and deck structure, etc., with no accommodation, services, utilities, mooring process, etc. The estimates assume a basic specification with typical compartmentation to meet MARPOL '78 and Classification Society requirements. The starting point for analytical techniques is broadly defined in ref.4.

Introduction - semi-submersibles
The Balmoral FPF is the only purpose-built unit yet installed. Two more are under construction, the SANA 15,000 and Petrobras XVIII. All three are based on modified drilling vessel designs where the marine systems layout and hull

structure have originally been orientated towards operating and moving between locations which may be very different. Obvious examples are the versatility and hence complexity, of the ballast system (including the compartmentation of hulls and columns) and the mooring system and associated equipment (including winches, chain lockers, etc.) The use of modified drilling vessel designs, therefore, represents a compromise on the initial configuration of an FPF unit which would spend most of its life at only one location.

If a purpose-designed and built semi-submersible is considered for a particular field development (or for more than one field, but with limited variables), considerable economies can be made as a result of optimisation of the design to suit its dedicated production role and particular location.

Although there have been many concept proposals for site-specific FPFs, there is limited constructional experience applicable to FPF semi-submersibles. Whereas the purpose-built monohull-based FPF can be based on existing commercial and technical solutions, the purpose-built, semi-submersible FPF base-case model needs to be considered in a different way. The approach adopted is based on similar parameters to TLP developments described in ref.5, and the following criteria.

(a) Deck optimisation for production, as opposed to drilling loads, e.g. no BOP handling, drilling derrick, etc.
(b) Deletion of unnecessary facilities, including adjustable moorings, active ballast system, etc.
(c) Rational assessment of the regulatory requirements as they would apply to a permanent installation, rather than a mobile drilling unit.

The shortage of actual data against which any theoretical model can be correlated is important when making comparisons with other proposed solutions. As the confidence level for base-case options using the semi-submersible configuration cannot be as high as for monohull or other structures, early consideration must be given to an independent verification of the design.

General selection

The hull selection process presented in this paper has been simplified to follow a single logical path which might be used to determine a typical hull type and size (schemes with combined drilling and production are, therefore, not addressed). The short example given is for the initial sizing of a monohull vessel in geographical areas where there is a limiting significant wave height of 10 m. A similar methodology is equally applicable to other FPF platforms within specific bounds of environmental and regulatory parameters.

Hull type selection
The selection of hull type is constrained by some key parameters.

Export method. If the facility is close to an existing pipeline, it may be possible to negotiate export via the pipeline. In this case tanker export will not be required, and with it the need for storage options. If pipeline export is available the choice would normally be a barge for milder environments and a semi for harsh environments. (Note: both options often need to be considered in anticipation of pipeline tariff discussions.

Well servicing. If the field requires either a large number of independent risers, or well work-over from the FPF, a non-weathervaning system is a first choice. When storage is not required then the semisubmersible is the first option in all but mild environments. A barge with non-weathervaning moorings can be used in very mild environments, but needs careful design development.

Discontinuous production. Shuttle tanker export, but with no storage. This is possible at the initial stage of a field development where there are no reservoir problems and high well productivity. Early North Sea fields were developed on this basis, but there will be fewer opportunities in future. The semi-submersible remains the first choice for the North Sea, with a flat-topped barge being the alternative in milder environments.

Mooring constraints. These have been a major factor, especially in the North Sea, because the initial developments have been in relatively shallow water. For any system requiring storage, and either a large number of wells or work-over, the mooring will have a major influence on the hull design. In harsh environments care must also be taken to ensure that the barge option, if chosen, meets mooring requirements.

Parameters for sizing the hull
The hull is sized initially by considering first the deck area requirement, and then the payload requirement. The structural arrangement of the FPF can be divided into five groups

(a) process
(b) utilities
(c) accommodation

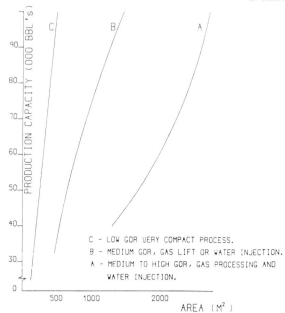

Fig. 1. Typical process area for various basic process plants

Table 1. Typical deck area requirements (m^2)

VESSEL TYPE	ACCOMMODATION	MOORING
Storage Barge (M)	640	480
Storage Barge (L)	840	765
Semi-submersible	625	NONE
Barge (no storage) Beam in meters 20 30 40	400 600 800	225 435 700

Table 2. Typical areas of additional items (m^2)

EQUIPMENT PACKAGES	MINIMUM SCHEME	UP TO 60,000 BPD	OVER 60,000 BPD
TEST SEPARATOR	-	100	150
GROUND FLARE	25	130	200
WI FINE FILTRATION	-	70	200
WELL SERVICES	50	75	150
FISCAL MEETERING	50	100	200
ACCESS AREA	20%	27%	35%

NORTH SEA INNOVATIONS AND ECONOMICS

(d) access
(e) mooring

If the FPF is to have storage then in addition to these groups the weight of the oil and ballast must be calculated for storage solutions (ballast volume can be taken as about 20% of the storage volume in the first estimate). The volume required can be determined from an examination of the export system, where the storage capacity is dependent on three variables

(a) production rate
(b) shuttle tanker capacity
(c) length of time tankers are away from field (this is related to journey distance, tanker speed, number of tankers, time spent unloading, time spent waiting on weather).

A more detailed analysis of the storage requirement should be made at an early stage as additional storage is expensive, and the storage volume can be sensitive to small changes in these variables.

Sizing the hull

Hull without storage (selected within constraints from above)

(a) Enter Fig. 1 at the production capacity and process type, and select an area, or develop a process layout plan area.
(b) Obtain first estimates of area for accommodation, mooring and access from Tables 1 and 2, or develop general area requirements.
(c) Enter Fig. 2 with the deck area to obtain the hull's dimensions, and displacement.
(d) Check that the load-carrying capacity of the selected hull is sufficient to carry the loads. This is more likely to be a problem for the semi-submersible than for the monohull.
(e) Once a suitably sized hull has been selected, read off the steel weight for costing purposes.

There can be great variations in the general dimensions of flat-topped barges, and the basic level of compartmentation. However the dimensions and proportions proposed give a reasonable base case on which to develop refinements.

Hull with storage

(a) As above.
(b) As above.
(c) Enter Fig. 3 with the deck area to obtain the hull's minimum dimensions

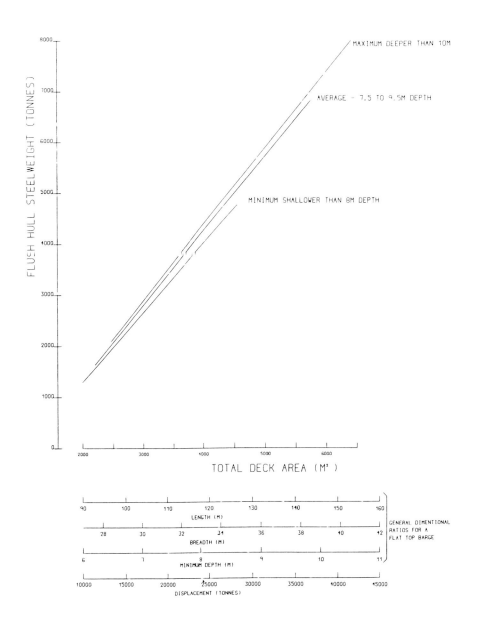

Fig. 2. Mono hull - no storage: flush hull steelweight without deck or mooring point

NORTH SEA INNOVATIONS AND ECONOMICS

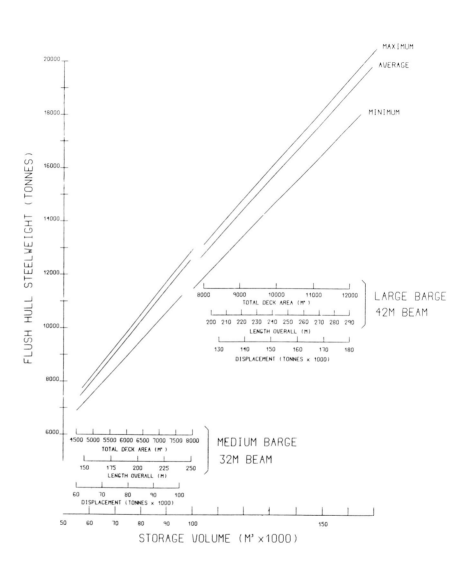

Fig. 3. Mono hull with storage: flush hull steelweight

Table 3. Additional specification features

SPECIAL REQUIREMENTS	ADDITIONAL FEATURES	EXTRA WEIGHT
Very heavy process	Extra deck steel	+ 1-3%
Extended life	Plate thickness	+ 2-5%
Difficult cargo	External stiffening	+ 3-5%
Tank separation	Extra compartments	+ 5%
Pollution control	Double sides	+ 5%
Pollution control	Full double hull	+ 15%

and storage volume.

(d) If the storage volume is less than the first estimate then the storage should be optimised as described, and this volume is then the limiting criteria.

(e) As above.

With the basic hull steel weight now estimated, additional specification features can be incorporated if required. Table 3 gives some examples of likely requirements.

Monohull moorings

Introduction
A wide variety of mooring systems have been developed, which have not generally been optimised for the particular applications. It is, therefore, difficult to use past project data to directly select the best possible system at the feasibility stage of a project. There are also quite a few cases where there is a possibility of choosing several systems, especially in shallow water.

For active (full Dynamic Positioning or Thruster Assist) systems there is a further complication of how different national governmental codes influence the design. The selection of the mooring arrangement should, therefore, be seen as an iterative process with the hull selection for all but the simplest systems.

Selection of system
The minimum number of parameters that are required for the selection of any mooring system are

(a) water depth
(b) wave height

(c) hull displacement
(d) number of individual risers.

To start the selection process, Chain Catenary Systems (CCS) with anchors have the widest range of applications. CCS systems should, therefore, be used for the first stage of feasibility analysis. The possible wide range of these preliminary parameters necessitates a further subdivision of the CCS for first analysis purposes. The subdivisions initially proposed are

(a) Conventional Spread Mooring (CSM) - where the FPF is moored on a fixed heading, with chain catenaries being directly attached to the fore and aft ends of the hull.
(b) Turret Mooring (TM) - the chain catenaries are brought together at one point, the turret, which enables the FPF to weathervane. Turret installations are further subdivided into light and heavy duty systems, depending on environmental parameters.

The initial selection of the system can be made by plotting the calculated parameters on Fig. 4, and selecting the basic type of mooring system. The first choice of mooring systems are shown in Figs 5(a)-(d). The following notes are given to check that the parameters are within practical boundaries and to assist in rating the relative merits of each system at the feasibility stage.

Figure 5(a): The simplest and cheapest option if WD and WH allow. Side by side offloading is difficult, so tandem loading or a separate export terminal. Usually makes the riser problems simpler. Fixed heading and position could permit limited well work-over in certain water depths.

Figure 5(b): Easiest system to add on to the hull, so now becoming the standard 'first choice' for mild environments. More expensive than small internal turret if this is designed into the hull at the concept stage.

Figure 5(c): Best technical solution if water depth and design of the hull permit easy installation.

Figure 5(d): High cost, and only required where there are large mooring forces, multiple risers and/or necessity to provide well work-over.

In circumstances where a simple CCS system is not applicable, a more detailed approach is required to select mooring systems which have additional elements of complexity. Such systems are usually required in shallow water (WD/WH) and alternative systems are shown in Figs 6(a)-(d). The following notes are given to check the practical boundaries and relative merits of each system.

Figure 6(a): Cost effective system in WD above 50m, with low environmental forces. More applicable to a storage barge that can be ballasted to bring the bow out of the water. More applicable with less than three risers.

Figure 6(b): Required in shallow water to make turret catenary

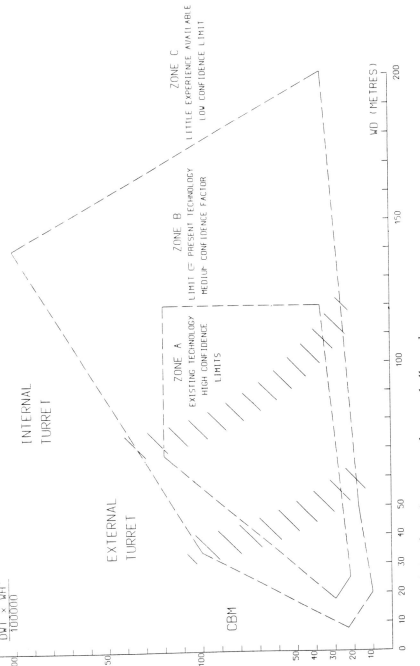

Fig. 4. First choice of mooring systems for mono hull vessels

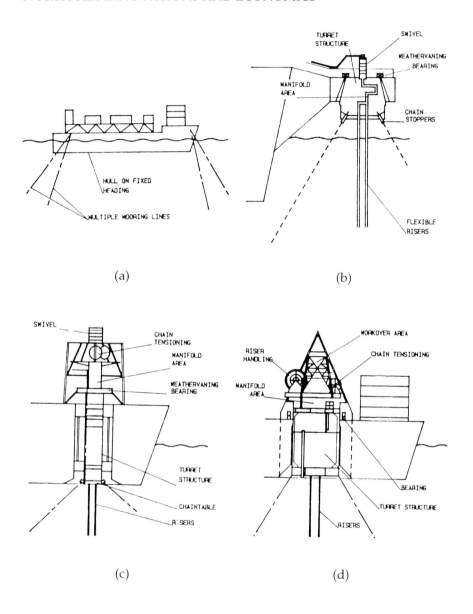

Fig 5. Simple moorings: (a) conventional spread mooring; (b) external turret mooring; (c) small internal turret; (d) large internal turret

system work, especially with larger vessels. More applicable with less than three risers.

Figure 6(c): For shallow water with high environmental loads. Alternative to system No. 6. More applicable to multiple -risers than No. 6. Only applicable for low environmental forces.

Figure 6(d): Applicable in shallow water with large vessels. Applicable for large number of risers.

Mooring - semi-submersible

A great deal of experience has been acquired in the past fifteen years regarding the behaviour of moored semi-submersibles and consequently design methods have been refined to quite a high degree. In the simplest terms the design has to accommodate the heave motions of the hull (pitch and yaw should be minimal) whilst minimising the horizontal motions as follows

(a) mean excursions due to the composite forces imposed by wind, waves and current;
(b) low frequency oscillations near the natural frequency of the hull and its mooring system;
(c) high frequency oscillations of the hull and mooring components associated with forces at wave frequency.

Design

The compliance of the mooring system must cater for all the horizontal excursions without excessive forces being generated by the wave frequency motions of the hull. Simple chain (or chain and wire) catenary systems can be developed up to 500m. For deeper water the design problem is to develop a system sufficiently stiff (with damping at below wave frequency) to maintain the platform on station, but with softness (with low damping) at wave frequency to absorb high frequency oscillations without creating excessive peak loads.

The design of the mooring system can be seen to interact with the hull configuration, water depth, excursion limitations for riser requirements and environmental parameters in a complex way. The first proposal is, therefore, to use a conventional catenary scheme which should be able to withstand survival conditions with the flexible production risers connected, and with both risers and the mooring system remaining intact and undamaged.

Selection

Designers and suppliers have acquired a great deal of knowledge in the area of fatigue phenomena, construction methods and installation procedures.

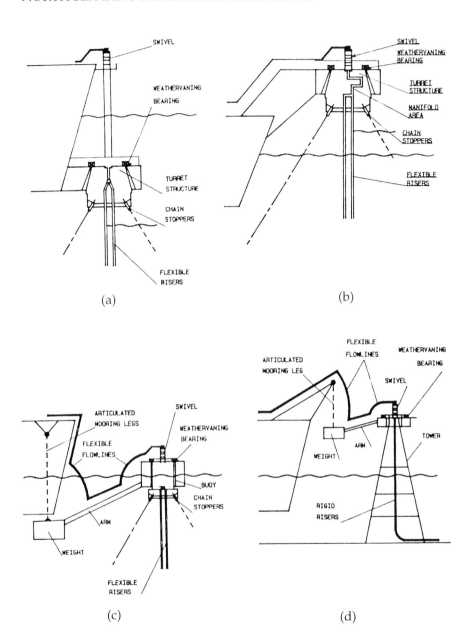

Fig. 6. Complex moorings: (a) external (submerged) turret mooring; (b) external (elevated) turret mooring; (c) yoke (soft) mooring; (d) tower (yoke) mooring

This makes it perfectly feasible to permanently moor a unit in up to 500m, in environments such as the North Sea. However, there are now so many Governmental Regulations, Industrial Design Codes and Classification Requirements that it is difficult to arrange a rational design route at an early stage in the project. We therefore recommend developing the approach suggested in ref.5 for the North Sea.

Conclusions

To identify development options at the earliest possible stage in a project development it is important to have available a reliable tool which can be used with limited data. A rational approach to the selection and sizing of FPF hulls and moorings can be achieved at the concept and feasibility stages without resorting to detailed engineering. To adopt this approach the development team need to have a reasonable model of the FPF development options, which also allows for rapid review of alternatives and good sensitivity analysis. This model should be supported by reliable empirical data, which give acceptably accurate base case schemes.

References

1. MOAN T. AND RINGSTROM B. Development of Rational Structural Safety Criteria for Offshore Structures with emphasis on ships. 6th International Conference on Floating Production Systems, London, December, 1990.
2. ROWAN D. Optimised Mooring Machinery for FPVs. 6th International Conference on Floating Production Systems, London, December, 1990.
3. MANOUDAKIS C. P., WILLIAMS L. M. and GRECCO M. G. A Guide to Cost-Effective Tanker-Based Floating Production Systems. Offshore Technology Conference 1987, No. 5489.
4. GRANT R. G., SIRCAR S. and NIKODYM L. A. A Systematic Procedure for Developing Optimum TLP Configurations. Offshore Technology Conference, 1991, No. 6570.
5. NYGAARD C. Concrete Floaters for Production and Oil Storage. 6th International Conference on Floating Production Systems, London, December, 1990.

Discussion

The speaker was asked if the smaller facilities pose limitations. In response the speaker said that the vessel motion in the smaller facilities were an important factor in the motion and mooring of such facilities.

NORTH SEA INNOVATIONS AND ECONOMICS

A delegate commented that the situation following the grounding of the Braer tanker in the Shetlands comes under the control of the Department of Trade & Industry. The NSE covers spillages from FPF. He said that it would be interesting to see how such topics would be dealt with in safety cases. The speaker added that the events following the grounding of the Braer could mean that it may become less favourable to use converted tankers in the future.

Another delegate enquired as to what governed the choice of use of FPF given that the driving force must be the field life. The speaker said that the perception of the risk involved was high as a result of the issues involved not being understood. However, the best return of any North Sea project was achieved by the use of a floater by Hamilton on the Argyll/Duncan/Innes fields. Also, demobilisation costs are low for a floater if the reservoir proves to be smaller than expected, as at Crawford.

Gannet's experience in reducing costs

J. H. T. CARTER, Shell Exploration and Production, UK

Introduction
Gannet is an oil and gas development in the Central North Sea. It consists of four separate fields and is located some 180km (112 miles) due East of Aberdeen.

The project is the latest in a long line undertaken by Shell UK Exploration and Production (Shell Expro) on behalf of Shell and its co-venturer Esso. As with many of its predecessors, it takes its name from seabirds indigenous to the region.

Discovery
Several hydrocarbon accumulations were discovered in the Central area of the North Sea in the period 1973 - 1984. Five of these discoveries (previously known as the Gannet Cluster fields and including Kittiwake) were under development at the time of the 1986 oil price collapse, as a result of which that project was cancelled.

Following a complete redesign based on a new 'simplified facility' concept, Kittiwake was reinstated as a stand-alone development in 1987 and is now operational. Adopting a similar concept, three of the remaining four original Gannet fields, together with a further discovery made in 1987, were identified as a potentially attractive grouping.

During 1987/88, feasibility studies demonstrated the economic viability of an oil and gas development centred on a production platform at Gannet A with subsea satellites at Gannet B, C and D and in mid-1989 approval was received from the Department of Energy for a development based on this proposal.

The Gannet fields contain an estimated 170 million recoverable barrels of oil and condensate and 704 billion cubic feet of gas.

The reservoirs
The four reservoirs which form the Gannet fields are good quality turbitic sands of Tertiary age, with good porosities and permeabilities. The oil and

Table 1. Gannet field

Accumulation	Reservoir	Geological description of trap
Gannet A	Tay	Structural/Stratigraphic
Gannet B	Tay and Forties	Pierced salt dome
Gannet C	Forties	Pierced salt dome
Gannet D	Tay, Rogaland, Andrew	Structural dome

gas accumulations are shallow, varying from 1,768 metres (95,800 feet) to 2,227 metres (7,305 feet) in Gannet A, B and C, while Gannet D is somewhat deeper at 2,468 metres (8,100 feet) to 2,728 metres (8,950 feet). All reservoirs are at hydrostatic pressure. Sufficient aquifer support is available in each field, and hence water injection is not required for pressure maintenance. However, gas lift will be necessary to enhance oil recovery from Gannet A, C and D, and compression flexibility is also available for re-injecting excess associated gas production to match gas sales requirements.

Gannet development plan

In order to ensure viability, the development plan placed heavy emphasis on minimum life-cycle cost associated with the highest achievable safety standards. It therefore utilised, where possible, existing infrastructure, simplified facilities and other cost-saving features.

The development plan was based upon a centrally placed fixed production platform at Gannet A, with subsea satellites at Gannet B, C and D. The platform is processing the oil and gas from all four reservoirs.

The Gannet A platform was designed around the concepts of simplified facilities based, where possible, on those proven in other recent developments and minimum manning requirements during operations. To assist this approach the project incorporates Tender Assisted Drilling (TAD), the first time that Shell Expro has used this concept.

Gannet B, C and D will be linked by pipelines to the central platform and will constitute the largest subsea development of any field in the UK sector of the North Sea.

Production commenced on 29th October 1992. Oil is being exported to shore via the Fulmar Floating Storage Unit (FSU) for offshore loading by tanker. Produced gas is exported via a spur to the existing Fulmar gas pipeline to the St. Fergus gas plant.

The field life is expected to be 20 years, with peak production of 50,000 bbl/day of oil, and gas compression capacity of 180 MM scf/day for both export and gas lift.

Cost-saving features

After the oil price collapse in 1986 it was clearly necessary to re-examine the Gannet project to find a more economic method of development. It was quickly realised that the revised development would only be viable if lifetime costs, i.e. both capital and operating costs, could be contained.

One approach was to use the concept of simplified facilities; this concept involves keeping the design as simple as possible, consistent with operational and safety objectives and utilising low maintenance equipment and materials.

The result is a simpler platform which is easier and safer to operate, with a significantly lower manning level than for any previous development. The other cost-saving approaches adopted were: extensive use of subsea development in preference to wellhead jackets; a large integrated deck containing nearly all the topsides facilities reducing installation and offshore hook-up; tender assisted drilling - drilling support only when needed; a lift-installed jacket - saving weight and fabrication costs; evacuation via the existing Fulmar facilities - utilising surplus capacity; minimum manning requirements during operations - reducing the size and facilities required for living quarters.

Safety

A feature of the Gannet development plan was to maximise the safety of personnel, the installation and the environment. This has been achieved by giving top priority to safety considerations at every stage of the design. Some of the main results of this 'designed-in' approach are as follows.

(a) The minimum manning philosophy - reducing the number of personnel at risk; the simple facilities philosophy and extensive use of low maintenance materials has resulted in a very small 40 person living quarters.
(b) Maximum physical separation of living quarters from the more hazardous areas of the platform. This is achieved by means of a buffer zone containing the utility systems enclosed by two blast walls.
(c) Extensive protection of critical steelwork utilising fireproof insulation.
(d) Enhanced fire and gas detection and protection systems.
(e) The installation of three fully enclosed lifeboats, each of which is capable of taking the entire operations crew.
(f) The provision of three separate fire pumps giving firewater availability

in excess of regulations.

(g) The fitting and careful positioning of emergency shutdown valves to prevent flow in the event of an incident.

Although the design of Gannet was being undertaken concurrent with the inquiry into the Piper Alpha disaster, almost all the recommendations in the Cullen Report had been anticipated and very few changes were required as a result. This serves to underline the project team's total commitment to providing the safest possible development.

Gannet A

Topsides

Gannet A topsides comprise an integrated deck with separately installed living quarters/helideck, drilling equipment set and flare boom. The operating weight of the topsides is estimated at some 16,000 tonnes, with the lift weight of the integrated deck, at 9,600 tonnes, being one of the heaviest offshore lifts ever undertaken. The deck structure contains the process, utility equipment, wellheads and control room.

Extensive use was made of computer-aided design throughout the design, fabrication and hook-up of the topsides.

Oil and gas from the B, C and D fields is piped to the platform, where process facilities provide each of the four fields with dedicated first stage separators and metering facilities for the separator products. These feed into commingled single train systems for oil and gas. Hydrocarbon streams leaving the platform are metered to fiscal standards.

Initial oil and gas production commenced from the subsea developments, whilst drilling of platform wells continues in parallel. At platform start-up a total of 6 oil wells and 2 gas wells had been pre-drilled and completed on these subsea fields. The Tender Assisted Drilling adopted for the platform wells entails using a semi-submersible vessel moored adjacent to the platform and connected to it by a bridge link and a system of catenary hoses and cables. The tender provides storage, mixing and pumping facilities for mud and cement as well as power and utilities. It also provides living quarters for the drilling crew.

This reduces the equipment located on the platform which is only required for the relatively short drilling phase, and therefore overall maintenance costs. The drilling programme at Gannet A is expected to take approximately two and a half years.

Substructure

The substructure is a four-leg lift-installed steel jacket standing in a water depth of 95 metres (310 feet). Fabricated on specially constructed skid shoes enabling it to be skidded onto a floating barge, the jacket had a lift weight of 8,400 tonnes. Each leg has 3 vertical piles of 2438mm diameter, the largest installed to date by Shell Expro. Pile penetration is approximately 85m due to poor strength soils in the upper 40m of the soil column. Six stab-in points support the integrated deck.

In addition to the loads associated with the 100 year return period storm, the jacket is designed to support a topsides load of some 16,000 tonnes and a total of 39 pipeline risers. A feature of the design is that 32 of the risers are contained within 3 sleeved bundles.

After transportation to the installation site, the jacket was installed with the two cranes of the Heeremac DB102 semi-submersible derrick barge. The lift and up-end was a single operation with no re-rigging of slings or removal of buoyancy tanks before setting on the sealed. This represented the heaviest offshore jacket lift carried out to date by Shell Expro.

Hook up and Commissioning

The final phase of the project covered offshore hook-up and commissioning of the platform facilities and satellite fields.

This phase of the project was planned to be shorter than previous Shell Expro projects of this size, because the prime consideration was being given during design to maximising the onshore hook-up and commissioning workscope, with a subsequent reduction in offshore hook-up requirements.

The satellite reservoirs

Gannet B is a gas field 5km north-west of Gannet A. Initially it has two subsea wells producing via individual 150mm pipelines to Gannet A.

Gannet C, 5 km to the south-west of Gannet A, has a thin oil rim and a gas cap. To develop it, 7 horizontal oil wells and 2 gas wells are required, each serviced by 100mm pipelines to Gannet A.

Gannet D is an oil field 16 km north-each of the platform. Currently it has four wells and oil will be transported to Gannet A through a subsea manifold and two 150mm pipelines.

Pipeline installation

All the conventional pipelines were installed using the pipe-lay vessel Lorelay. The process involved on-deck welding of the individual lengths of coated pipe and continuous laying across the stinger at the vessel's stern. The B and D in-field pipelines were trenched and rock dumped to prevent upheaval buckling.

The Gannet bundles are in-field pipelines which connect the subsea wells

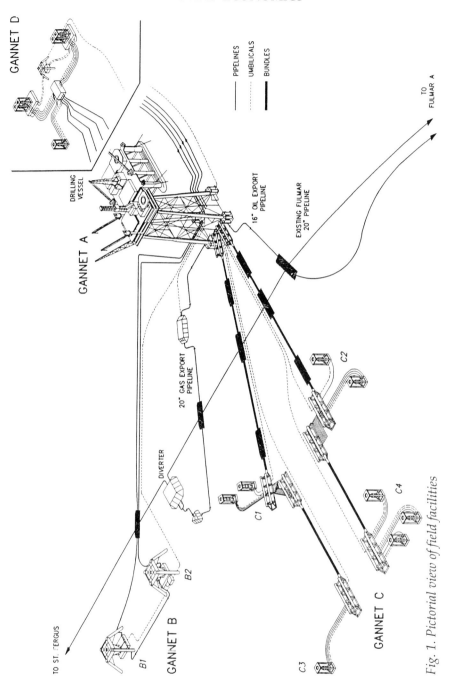

Fig. 1. Pictorial view of field facilities

at Gannet C to the platform at Gannet A.

Bundles were selected for this application as the best engineering and economic option, given the number and complexity of pipelines required to service this field.

Two Type A bundles (one 3.6 km long and one 3.2 km), which connect the platform to the 2 nearest drilling centres, C1 and C2, are 946mm in diameter. Each contains fourteen individual 50mm, 100mm and 150mm lines. The remaining two Type B bundles (one 2.6 km and one 2.2 km) connecting the nearside drilling centres to the two outer drilling centres, C3 and C4, are 746mm in diameter and each contains eight 50mm, 100mm and 150mm pipelines. The thickness and diameter of the outer of carrier pipe was precisely designed to give a predetermined level of buoyancy for the completed bundle, allowing them to be towed out from the onshore fabrication site to their final location offshore by the mid-depth-tow method. Production conditions at Gannet C required the pipelines to be insulated by utilising a glycol-based gel injected into the carrier pipe annulus after installation. Each bundle, therefore, also incorporates a perforated 175mm pipe to accommodate expansion of this gel.

Final installation of the bundles involved docking them under Gannet A, using a specially installed temporary winch deck on top of the jacket.

Oil export
Oil is exported via a new 400mm diameter 107 km pipeline to the Fulmar A platform and then through the existing export facilities to the Fulmar Floating Storage Unit (FSU). The FSU has storage facilities to enable production to be maintained continuously. Oil is off-loaded at periodic intervals into shuttle tankers which transport the crude ashore.

Gas export
Gas, including natural gas liquids (NGLs), is piped via a new 500mm 1.8 km long pipeline into the existing diverter of the Fulmar gas pipeline at Gannet.

From this diverter, the gas is transported via the existing Fulmar gas pipeline to the onshore gas plant at St. Fergus, where the natural gas liquids are separated out and dry gas is made available for sale. The remaining natural gas liquids are transported by an overland pipeline to the Mossmorran fractionation plant in Fife, where they are broken down into their constituent parts and sold.

Conclusion
The development of the Gannet fields can be seen as a major milestone in

reducing costs in the North Sea. In particular, if we are to face the challenge of developing safely the smaller hydrocarbon accumulations in the future we will need to build on the Gannet experience.

Discussion

A delegate asked whether free-fall lifeboats were used on Gannet. In reply the speaker said that the conventional lifeboat system was used. However, Shell have adopted free-fall lifeboats on Nelson.

Another delegate wished to know if the tender vessel was moored. The speaker replied that it is moored using anchor piles but has power available to pull away should it need to do so in bad weather.

The speaker was then asked to expand on comments about a lower lifetime cost. The speaker replied that designing a minimum facilities and minimum manning platform led to a low CAPEX. However Gannet was also based on minimum maintenance, which was achieved for example by using exotic steels in the pipework of the topside. Shell had also tried to ensure that any decision on material/equipment looked at a lesser cost in the operating phase. These two mechanisms led to a low OPEX.

Economics of pipeline bundles

A. PALMER, Andrew Palmer and Associates Limited, UK

Introduction
Satellite developments often require several pipelines between the same end points, for different functions such as oil production, gas production, water injection, test, annulus access, and chemical injection. They also require control umbilicals, electrical power cables, electrical and fibre-optic telecommunications cables, and heat tracing pipes or cables, again between the same end points.

All these lines and cables could be laid individually, and there is a well-established technology for doing this. A clear alternative is to lay the lines and cables together, in a single bundle. This will sometimes be significantly cheaper. The lines can simply be strapped together, in an 'open' bundle, but more often they are contained within an outer carrier pipe, as a 'closed' bundle. The carrier pipe then has several functions.

The carrier protects the lines and cables mechanically, so that they are secure against mechanical damage during construction and subsequent operation; in particular, if the carrier diameter is 16 inches or more, experience shows that it is safe against external damage from North Sea fishing gear and does not need to be trenched.

The carrier helps to control the submerged weight of the bundle, particularly in construction methods such as mid-depth tow, where precise weight control is critical; the annular space between the internal lines and the carrier can be air-filled or nitrogen-filled during construction, and flooded with some heavier fluid for enhanced stability during operation

Within the carrier, a controlled corrosion environment for the internal lines and cables exists, and this makes it possible to eliminate or reduce requirements on the anti-corrosion external coatings of individual lines: if the annulus is left nitrogen-filled, no anti-corrosion coating is required, and in any case the anti-corrosion coating does not have to have mechanical strength to resist damage during operation.

Within the carrier, a controlled thermal environment for the internal lines and cables exists; this widens the options for thermal insulation, and adds new options such as filling the annulus with a gel.

Within the carrier, there is a better environment for umbilicals and electrical and fibre-optic cables, so that they are secure against mechanical damage (so that armouring will not be required, for instance), and they will be protected against water if the annulus is dry.

All of these factors reduce the costs of the internal flowlines and of the cables.

These are advantages of a bundle in a carrier, but there are disadvantages too. The carrier itself adds significantly to the steel and fabrication costs: usually, there is more steel in the carrier than in all the internal lines put together. The presence of the carrier makes the bundle much stiffer is flexure: this is sometimes an advantage but may be a disadvantage if (for instance), the bundle is to be pulled into a J tube or an I tube. Finally, although the internal lines are well protected and less likely to be damaged, if one of them does begin to leak it is relatively difficult (though not impossible) to locate the leak and to get through the carrier into the bundle to carry out a repair.

Bundle technology is by no means new. For instance, a flowline bundle was installed for Panarctic at Melville Island in the Canadian Arctic as long ago as 1978 (ref. 1,2). It connected a single subsea gas well, Drake F-76, to the shore 1100 m away, as part of a project to demonstrate the feasibility of producing Arctic offshore hydrocarbon resources under sea ice present through almost twelve months of the year. The technology has come a long way since then.

The overall cost reductions that follow from a bundle concept can be substantial, but each case has to be examined individually. The design of pipeline bundles is in fact in some ways more straightforward than the design of individual lines, because different parts of the system have different structural functions and there are fewer conflicting requirements.

A systematic approach is to design the internal lines first, using accepted principles of design for hydraulics, stress and corrosion. If the bundle is an open strapped bundle, the whole system is then checked for stability, for security against mechanical damage, and for compatibility with the installation and trenching method. If the bundle is a closed bundle in a carrier, the next step is to design the carrier, which must:

(a) be large enough to accommodate the internal lines, without crowding that would make fabrication difficult;
(b) have enough wall thickness for mechanical protection, and for strength during installation;
(c) give the bundle a submerged weight compatible both with installation and with stability during operation;
(d) be secure against external and internal corrosion.

Laybarge construction is often applied to the simplest form of bundle, a single large diameter line with one or two much smaller lines strapped to it (piggybacked), and occasionally to two medium diameter lines if parallel firing lines can be set up on the barge. Laybarge construction of more complex bundles is extremely slow and time-consuming, at least without extensive modification of the barge, and this is unlikely to be economic for a single project.

Tow installation of bundles

Bundles are generally made on shore and installed by tow. The versions of tow construction applied in practice have been mid-depth tow, controlled-depth tow, dynamic towing and bottom tow, though other tow options are possible, among them surface tow and near-surface tow. There have been some twenty-five North Sea bundle installations to date, and all of them have been mid-depth tows, from two make-up sites in northern Scotland. Bottom tow has been applied in other parts of the world, such as the Australian Bass Strait and the Gulf of Mexico.

Mid-depth tow carries the bundle suspended between two tugs, in a long catenary. With a bundle length of 6000 m and a water depth of 50 m during tow out, the catenary has to be extremely flat, and requires a very low controlled submerged weight, of the order of 25 N/m (2.5 kg/m), to keep the bundle clear of the seabed without the need for excessive tension from the tugs. The bundle is first launched into shallow water, where its weight is adjusted by trimming chains: this is a critical but tricky and sometimes time-consuming operation. When the bundle is under tow, hydrodynamic lift on the chains helps to reduce the effective submerged weight and flatten the catenary. The chains also have a secondary role. It was initially feared that a long towed flexible bundle might be liable to severe lateral oscillations, of the kind that afflict flexible tubes towed on the surface, but this seems not to happen: a probable explanation is the high degree of damping introduced by the chains.

On arrival at the installation location, the bundle is lowered to the seabed. The ends float above the bottom of the sea, lifted by additional buoyancy tanks and held down by drag chains, and they can be flexed laterally into alignment with the structures they are connected to. Almost invariably, the bundle needs to have additional weight if it is to be stable against hydrodynamic forces on the seabed, and this is secured by flooding the annulus, generally with water, but in the recent Gannet project this was done with insulating gel.

Bundles installed by bottom tow have been open bundles without a carrier. The submerged weight again has to be controlled accurately, so that the frictional against the seabed is not too large for the tug to overcome. The pipes (or the carrier if there is one) have to have anti-corrosion coatings rugged enough to resist the abrasion of the seabed during a long tow, and this is an extremely demanding requirement.

Reel installation of bundles

A completely different alternative is to make up the bundle onshore, to wind it onto a reel, and to lay it in place, following established reelship technology. A major advantage of reeled bundles is that there is no constraint on the route, which can incorporate fairly sharp curves, whereas towed bundles have to be nearly straight, except possibly at the ends.

The simplest application of the reeled bundle concept is the pipe-in-pipe concept, in which a flowline is enclosed in an outer pipe. The annular space can be filled with thermal insulation. This concept was applied in the Seahorse and Tarwhine fields in the Bass Strait (ref.3). It was initially thought that bulkheads would be needed at frequent intervals to suppress the development of large axial forces in the internal line, but that fear was in fact unfounded.

Reeling can be extended to bundles consisting of multiple flowlines (ref.4). The simplest version is an open bundle consisting of two flowlines with the same outside diameter, reeled side-by-side. The flowlines undergo the same deformations during the reeling, unreeling and straightening operations, and there is no tendency for the bundle to twist.

The next option is to have two flowlines side-by-side inside a larger carrier pipe, arranged so that the centre of each line is on the neutral axis of bending for the carrier, the diameter perpendicular to the plane of bending on the reel. On the largest reel system currently available, the reelship Stena Apache, the largest pipe reeled to date is 16 inches OD (406.4 mm). Taking the carrier wall thickness as 17.46 mm, giving a D/t ratio of 23.3, the inside diameter of the carrier is 14.62 inches (371.5 mm), which allows for two nominal 6 inch (168.3 mm OD) flowlines.

If there are three pipes in the bundle, the two obvious possibilities are: again to set all the centres in one line; or to arrange the cross-section as a compact bundle, so that each pipe is in contact with the other two.

The advantage of arrangement 2 is that it makes the bundle more compact, and the carrier can be smaller. The disadvantage is that the neutral axis of bending cannot go through all the centres, so that the when the bundle is bent at least one of the pipes is in compression. That pipe may then yield in compression, or it may buckle out of the bundle.

An alternative arrangement overcomes this. In a helical bundle, the pipes are wound helically around each other, or around a central core, like the strands of a wire rope, and allowed some freedom to move longitudinally.

The bundle can then curve relatively easily, since the additional length required for the tube to follow the outside of the curve is balanced by the reduced length required for the same tube to follow the inside of the curve, half a helix pitch further along. This is the principle which allows a wire rope repeatedly to be flexed to a radius of 25 rope diameters without low-cycle fatigue damage. In this instance, it allows a helical bundle to be formed while it remains in the elastic range, without plastic deformation.

A helical bundle responds to axial compression in a different way to a straight bundle. Each helix can deform by shortening axially and by increasing its diameter, so that the axial compressive strain and stress are relieved by additional bending deformation. This configuration reduces the risk of buckling. Its secondary advantages in countering upheaval buckling were identified in a SIPM joint industry study (ref.5). As often happens, it is far from a new idea, and was patented in the nineteenth century.

A feature of the reeled bundle in a carrier is that it eliminates trenching without losing the advantages of the reel method. Existing reel technology shows that pipe up to a maximum 16 inches in diameter can be reeled. By a happy chance, 16 inches is the minimum diameter that is agreed to be safe against North Sea fishing gear.

A major advantage of a reeled bundle over a towed bundle is that its installation is far less sensitive to submerged weight. The weight of towed bundles has to be controlled extremely precisely, which increases pipe costs and fabrication costs, and leads to costly delays while the bundle is trimmed in shallow water. If the weight is incorrect, the bundle will be difficult to control during the installation, and can either float up towards the surface or sink out of control towards the bottom. The requirement for minimum weight during tow conflicts with the need for on-bottom stability, and so weight has to be added once the bundle is in place.

In contrast, a bundle laid from a reelship only has to be light enough not to overload the tensioner, and this requirement is not difficult to meet in steep-ramp laying. There is no need to add weight by flooding the annulus, and this means that the annulus can be left nitrogen-filled or vacuum-filled, which eliminates the need for anti-corrosion coatings on the internal lines, and allows cheap foam insulation to be used.

The reeled bundle appears to be an extremely promising option for rapid and cost-effective construction.

Economies in design and construction
How can we reduce costs?
The first task is to reduce the costs of design activities. This can be done by

systematising the design process, and incorporating the calculations in user-friendly software, backed up by library design data and with straightforward outputs of documented design deliverables. Since the vast majority of design calculations are mathematically simple, the calculations can easily be carried out on a microcomputer. The software to accomplish the design process for single lines has been available for some time (ref.6), and makes it possible to carry out in a few hours designs that formerly took weeks, and to include a higher degree of design optimisation. Applications to bundles are now being developed. For instance, a bundle spreadsheet allows the designer to draw a bundle of the screen, while the program checks for clearances between lines, gives him or her a running output of bundle submerged weight and centre of gravity, and allows rapid editing. This allows the designer rapidly to optimise the cross-section, and then to go on to subsidiary calculations of heat transfer, on-bottom stability, buckling, spacer intervals and so on.

One of the largest cost components is the materials cost for the internal lines. This can be achieved by minimising the wall thickness. If the internal lines operate at high temperature, this can be achieved by adopting allowable strain design, nowadays allowed by the majority of codes (including the new BS8010 part 3). Allowable strain design (ref.7) keeps a restriction on allowable hoop stress, but drops the limitation on equivalent stress imposed by old codes such as IP6, and replaces it with a new condition on allowable strain, under conditions which are almost invariably satisfied. This change brings with it an automatic economy in materials costs.

The allowable strain option may lead to a substantial reduction in flexural stiffness (ref.5) at high operating temperatures and pressures, and this effect needs to be kept in mind in buckling calculations.

A second large cost component is the carrier. In a towed bundle, the diameter of the carrier is determined by buoyancy considerations, and the wall thickness by mechanical strength, in particular the requirement to avoid collapse under external pressure. The carrier wall thickness can be reduced by pressurising the annular space with nitrogen after the bundle has been completed and sealed, and possibly after launching, and this change in turn may make it possible to reduce the carrier diameter.

The bundle option already reduces the cost of thermal insulation, by eliminating the requirement for the insulation to be mechanically robust. If flooding can be avoided, the insulation does not need to withstand water pressure.

Further economies can be made in spacer design. Spacers have been made from various materials, among them cast aluminium and pressed and welded steel. Studies have shown that it maybe more economical to use polymers and elastomers, which are lighter, have low electrical and thermal conductivity, and avoid corrosion problems introduced by the presence of different metals. The quantities involved are relatively large, and the stresses relatively small: this suggests that it may be most economical to use cheap materials with adequate mechanical behaviour, such as filled polyurethane, recycled polymers and recycled rubber.

Installation costs can be further reduced by improving the productivity of make-up systems. It would clearly be possible to for improved production-line techniques to be applied to make-up, as was done in the World War 2 PLUTO project. A significant factor has been the uneven and uncertain flow of contracts, which has discouraged contractors from major investments in makeup sites.

Research and development needs

Bundle construction is nowadays an established technology, and design and construction can be carried through without additional research and development for a specific project. However, there are a number of outstanding open questions whose solution would allow further economies and give additional confidence to design

(a) is it necessary to provide internal lines with external anti-corrosion coating?
(b) can any leakage of water be tolerated, at the ends of the bundle or through carrier leaks?
(c) how can we detect and monitor leaks from the internal lines into the annulus?
(d) is internal galvanic action between carbon steel carrier and internal CRA lines important?
(e) would internal CP work ?
(f) if the annulus is flooded, are sulphate-reducing bacteria a problem, and if so, how can they be controlled?
(g) what is the optimal heat tracing system ?
(h) is there a risk of stray currents from power cables and heat tracing?
(i) does iron contamination and weld sputter on the external surfaces of duplex pipes matter in a deaerated seawater environment?
(j) if the annulus is flooded, would alternative fluids be advantageous?
(k) what is the optimal insulation system ?
(l) if gel insulation is used, is it necessary to provide for thermal expansion of the gel?

A few of these questions have been studied in the context of specific designs, by operators, contractors and consultants, but there has been little research into them.

One of the arguments against bundles is that no-one knows how to repair them. There is a need for research into both conventional techniques such as hyperbaric welding and unconventional techniques such as memory-metal sleeves, and into design to make repair more straightforward. Another objection to bundles in carriers is that the annular space between the internals and the carrier cannot be inspected by current technology. There is a need for research to find solutions to this problem.

References
1. PALMER A.C., BAUDAIS D.J. and MASTERSON, D.M. Design and installation of a submarine flowline in the Canadian Arctic Islands. Proceedings, Eleventh Annual Offshore Technology Conference, Houston, 1979, vol.2, 765-772.
2. MARCELLUS R.W. and PALMER A.C. Shore crossing techniques for Arctic submarine pipelines. Proceedings, Fifth International Conference on Port and Ocean Engineering under Arctic Conditions, Trondheim, 1979, vol.3, 201-215.
3. MCQUAGGE C.H. and DAVEY S. Bass Straits - an Australian Experience. Proceedings, Offshore Pipeline Technology Seminar, Copenhagen, 1991.
4. PALMER A.C. and HULLS K. Reeled pipeline bundles. Proceedings, Offshore Pipeline Technology Seminar, Amsterdam, 1993.
5. PALMER A.C., ELLINAS C.P., RICHARDS D.M. and GUIJT, J. Design of submarine pipelines against upheaval buckling. Proceedings, Twenty-second Offshore Technology Conference, Houston, 1990, vol.2, 551-560, OTC6335.
6. PLUS.ONE software documentation, 1991, Andrew Palmer and Associates.
7. PALMER A.C. Limit state design of pipelines and its incorporation in design codes. Proceedings, Design Criteria and Codes Symposium, 1991, Society of Naval Architects and Marine Engineers, Houston.

Discussion
The speaker was asked whether helical pipeline bundling had limited applications if all pipes in the bundle had to be of the same size. In response the speaker stated that what was shown was diagrammatic. Helical bundles can be made using differently sized pipes.

A delegate enquired about the knowledge of repair techniques. The speaker said in response that not a lot is known about what a repair to a bundle would be like. Obviously if the damage is to the carrier and external it would be easy to locate. However, internal damage would be difficult to trace, and would probably have to be traced acoustically. Another speaker added that if a line were to fail on Gannet, then the cheapest option would be to run a new line outside the carrier.

Elf Enterprise Caledonia Piper/Saltire pipeline bundler

T. W. TACONIS, Rockwater Limited, UK

Introduction

In 1980 the first flowline bundle was fabricated onshore and towed to the field using the controlled depth tow method (CDTM). In the past twelve years the CDTM and tow methods have become established and accepted techniques for the development of satellite fields. A logical new development is the full integration of the production systems and flowline bundle, which avoids separate installation of the production system.

In particular, subsea hook-up connection to the flowlines and umbilicals are no longer required. The integrated flowline bundle production system will be fabricated and fully tested onshore, before being towed offshore, which secures maximum reliability. The system offers maximum design flexibility and can be fine tuned to suit a particular field development plan. The system offers substantial cost savings and can be designed for water depth up to 2000m.

The bundle can contain all types of flowlines and umbilicals, whereas the production system can contain manifolding, templates, Underwater Safety Valves (USVs), control systems, multiphase pumping, subsea separation or other processing facilities. The production system can be at either end of the bundles or at one or more intermediate locations. The other end of the bundle can be fully integrated with a riser base or a direct pull-in structure. It can also be an onshore facility.

Recent experience on the longest and largest bundle of 6.7 km length and 40" diameter, which was successfully installed in the Piper/Saltire field for Elf Enterprises Caledonia in May 1992, will be highlighted. Both towheads with sizes of 23.5 x 4.4 x 5.5 m (length x width x height) and a total weight of 135 tonnes included 3 USVs.

Subsea production dates from the mid-sixties, when simple subsea trees were installed in shallow water depths and were possibly operated by divers. Ever since, the subsea multi-well manifolding and floating production technologies have been developed to provide total subsea concepts which now operate satisfactorily. However, these are generally in diver-accessible depths. The present research and development efforts of the oil majors aim for total remote production systems for developments in water depths be-

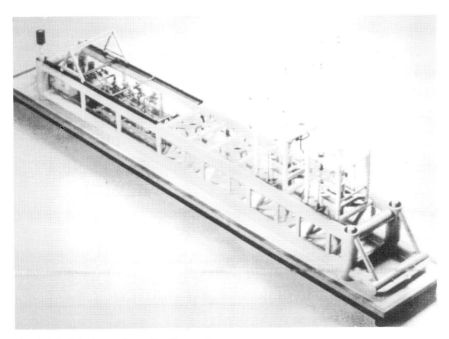

Fig. 1. Model of towed production system

yond diver access by the mid-1900s. Key objectives in the design of such a complex remote system are to achieve the highest possible system reliability and to develop efficient procedures for installation, commissioning, operation and subsea intervention.

The new technologies developed are complying with the above requirements, but will also be economic for shallow water application within diver depths.

A subsea satellite development typically involves a cluster of wells or a template structure and the connection flowlines and control lines to the parent facility. The paper introduces the concept of the Towed Production System (TPS) (see Fig. 1). The system combines all required flow and control lines within a bundled configuration, together with integrated facilities for manifolding and processing. It is based on onshore construction and testing of a total system which can be installed offshore by using a towing technique, thereby minimising underwater activity during construction and operation.

Installation of submarine pipelines by towing techniques is now an established and accepted method of installation. Pipelines can be installed by bottom tow (with or without added buoyancy), off bottom tow, near surface or surface tow, or using the Controlled Depths Tow Method (CDTM) (see Fig.

2). The CDTM is particularly suitable for transportation of integrated systems fabricated and tested onshore (TSP). This relatively new technique for bundle transportation is outlined in the following section.

General description of tow methods
Bottom tow
The simplest tow method is the bottom tow, whereby the bundle is in direct contact with the seabed, see Fig. 3. This requires an obstacle-free tow route and provision needs to be made to cross existing pipeline. For heavy bundles additional buoyancy may be required in the form of buoyancy tanks or a carrier pipe.

Off-bottom tow
With the off-bottom tow the bundle floats at a fixed distance several metres above the seabed, see Fig. 3. This is achieved by providing buoyancy, which is again offset with the weight of chains. The free hanging chain weight compensates for the buoyancy. The chain weight on the seabed determined

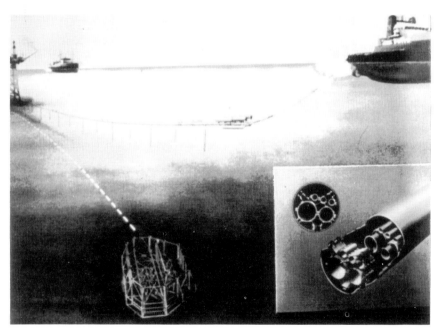

Fig. 2. Controlled depth tow method (CDTM)

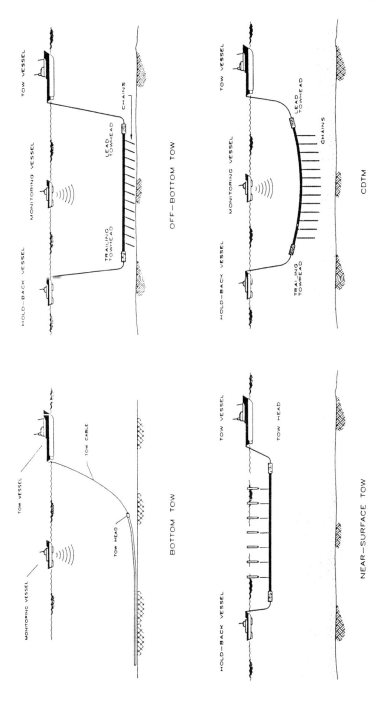

Fig. 3. Tow methods

the bundle's submerged weight. The advantages are the absence of seabed contact of the bundle, the lower tow forces, and the higher manoeuvrability.

Near-surface tow
With the near-surface tow the bundle floats at a constant depth below the surface, thus avoiding all seabed contact, see Fig. 3. The disadvantages are the influence of the waves and current and the fact that in deeper water controlled lowering of the bundle can be problematic.

Controlled depth tow method
The controlled depth tow method combines the advantages of the off-bottom tow and the near-surface tow. At low speed the CDTM is equal to the off-bottom tow. With gathering speed the increasing lift forces on the chains and the bundle reduce the submerged weight, and the bundle takes off from the seabed like an aeroplane. At very high speeds the bundle will come to the surface. Over the past twelve years a system has been developed to control the depth of the bundle by varying parameters such as tow force, bundle tension, and tow speed.

The bundle depth is controlled such that the wave influences are minimal, say 30m or more below the surface and well above the seabed say 30m or more, such that seabed obstacles can be safely passed. On arrival at location the bundle is landed on the seabed, again like an aeroplane, by reducing speed.

Piper/Saltire pipeline bundles

During the summer of 1992, Rockwater's Controlled Depth Tow Methods (CDTM) technique was used to install two pipeline bundles in the Piper field. The two bundles, one the largest as well as the longest ever towed in the CDTM mode, were installed to connect the Saltire A platform to the Piper B platform, and Saltire A to its satellite water injection template (see Fig. 4). At both Saltire and Piper ends of the longest bundle, subsea structures were required to house all the Underwater Shutdown Valves (USVs) for each individual line. Use of the CDTM technique allowed these structures to be incorporated onshore at both ends of the long bundle; this eliminated both separate installation of these structures and tie-ins between the pipelines and the USV structures.

Furthermore lift enhancement during tow was applied to enable towing the bundle in CDTM mode with reduced pull force requirements. Both the integration of USVs and the life enhancement technique are new features for the installation of pipeline bundles which will increase the potential applica-

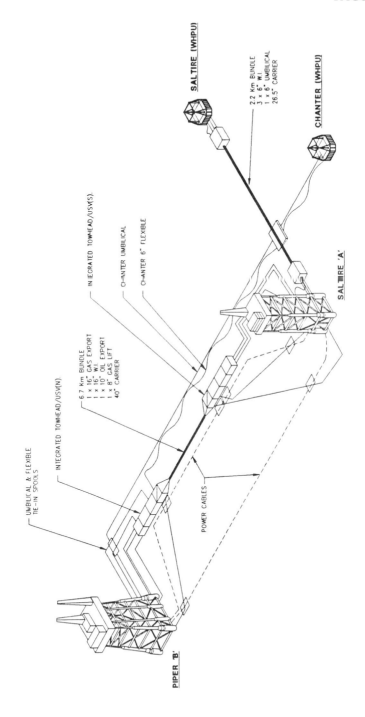

Fig. 4. Piper/Saltire pipeline bundles and power cable layout

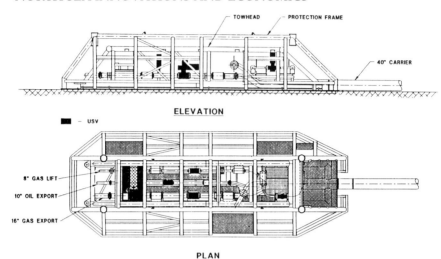

Fig. 5. Towhead/USV arrangement with post-installed protection frame

tion of bundle installation.

Integrated USVs

The larger 40" diameter bundle to connect Saltire A and Piper B contains a 16" gas export line, a 16" water injection line, a 10" oil export and an 8" gas lift line. The total length of this bundle is 6.7 km. Each end of this bundle comprises a towhead structure which contains the USVs and associated control systems, isolation valves, control umbilical termination housings and pig launching and retrieving facilities.

The Saltire end contains USVs for the 16" gas export line, the 10" oil export line and the 8" gas lift line. The Piper B towhead contains USVs for the same pipelines but, in addition, it houses the USV and associated duplex pipework and control system for the 6" flexible pipeline from Chanter, a subsea satellite field located to the South East of Saltire (see Fig. 5).

The smaller bundle connecting Saltire A to its dedicated water injection structure contains three 6" water injection lines and a 6" hydraulic umbilical. The bundle is 2.2 km in length.

Fabrication and launch

The pipeline bundles were fabricated at Rockwater's construction base at Wick in North East Scotland. As the longest bundle was longer than this facility, it was built in two sections, essentially 5.7 km and 1.0 km in length. The bundle sections were supported by bogies on two narrow gauge rail tracks. The two towhead/USV structures, each weighing 110 tonnes, were transported to site using trailers during ferry and road transport. Each towhead was also equipped with buoyancy tanks during the bundle tow, bringing the total weight of each structure up to 135 tonnes. The weight of the complete bundle is 5870 tonnes. The towheads were welded to the end of each bundle section by means of an end terminating bulkhead and a carrier pipe section. The inner pipelines are then hydrostatically pressure tested. The carrier annulus is filled with nitrogen, is pressure tested and remains filled with nitrogen for tow.

The USVs and control systems were checked prior to launch. The compartmented buoyancy tanks were partially filled to give the towheads neutral buoyancy during tow.

The 5.7 km section was launched initially and pulled out by a pull barge which moved on its anchors. The first section was fully launched and the landward end was clamped at the fabrication shop; the shorter section was then lined up and the two sections welded together.

The complete bundle was launched by the pull barge, and one of the leading tugs was attached to the bundle tow line. A second tug was then hooked to the bundle, providing two tugs at the leading end of the tow and the pull barge released. The third tug was located on the trailing end of the bundle to provide hold-back tension.

After completion of the launch the pipeline was floating 4-5m off bottom with the lower links of the ballast chains resting on the seabed. The number of links of each chain on the seabed determined the bundle submerged weight. Adjustment took place by removing part of the chain to achieve the desired submerged weight.

Lift enhancement during tow

Until the Piper redevelopment project the maximum length of a pipeline bundle installed using CDTM was 4.5 km. As the required tow force will increase with the bundle length, the conventional method with bare chains suspended from the bundle required improvement. Several options were investigated in an early stage of the project, resulting in wrapped chains, i.e. chains sheathed with a fibre reinforced PVC cover. The final choice of the wrapping material was based on abrasion, tear, strength, fitting and flexibility characteristics. Two materials were selected to test their hydrodynamic characteristics, i.e. lift and drag behaviour. The full scale tests were carried in the Deep Water Tank of MARIN, the Netherlands. During the tests the lift and the drag of bare chains and wrapped chains was determined as a function of the tow speed (see Fig. 6). After the tests the final wrapping material was selected. Comparison of the hydrodynamic characteristics of bare chains

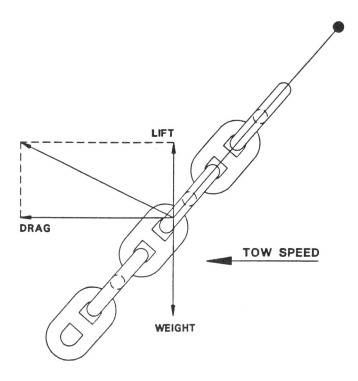

Fig. 6. Drag and lift forces on chains during tow

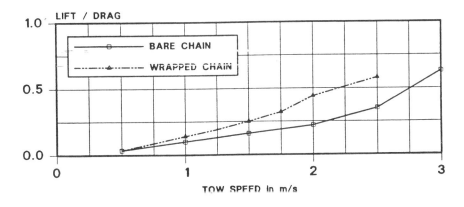

Fig. 7. Lift/drag ratio for bare chains versus wrapped chains

versus wrapped chains (see Fig. 7) showed that the lift over drag ratio improves considerably using the wrapping material. As a result, the required pull force was reduced by 17% while the required tow speed was reduced by 25%. Both reductions resulted in smaller leading tugs and lower stress levels in the bundle.

Tow and installation

Prior to tow, a vessel was deployed to survey the tow route and installation area, and set transponders to facilitate the installation and final positioning. The total length of the tow from Wick to the Piper field was 230 km.

By increasing the tow speed during tow initiation the chains suspended from the bundle started creating a lift effect so that the pipeline bundle became free from the seabed; the height of the pipeline bundle above the seabed can thus be controlled by the tow speed. The bundle is now suspended between tugs on the leading end and the trailing end (see Fig. 2). The depth of the bundle is monitored by means of transponders along the bundle which send their information acoustically to the tow master on board the command vessel. The tow master used this information to adjust the tow speed by small pull force variations to assure that the bundle is being towed at the required depth. Prior to the tow, calculations based on historical information and model tests are used to determine tow speed.

The bundles were positioned in their assigned corridors in the following manner. The tow of both bundles approached the Piper B platform from the North, passing it on the Eastern side. This put the bundle in line with the 10 m wide installation corridor. As the tugs slowed, the bundle gradually lowered to the off-bottom mode, floating 4-5 m above the seabed in a water depth of 145m.

The bundle was then slowly manoeuvred into its final position within the corridor, the bundle annulus was flooded with inhibited seawater. The towhead buoyancy tanks were then flooded and removed. To complete the project after the pipeline bundles were installed, the pipelines were pigged using pigs pre-installed onshore. This was followed by hydrostatic tests performed from a DSV. The subsea pipelines were connected to the platforms and water injection satellite by spoolpieces.

Further developments

The following further innovations were incorporated in the launch, tow and installation of the 40" diameter 6.7 km bundle due to length and weight consideration.

(a) Modified bogie design using steel wheels, reducing launch forces with a factor of 10 minimum.

(b) Launchway with rollers to support bundle and towheads on the beach section where the bogies are disengaged.

(c) Intermediate tie-in of two bundle sections during the launch.

(d) Transport of the bundle with two leading tugs using a bridle arrangement.

(e) Monitoring the bundle depth using two survey vessels linked with a telemetry system.

Towed production system

The concept of the Towed Production System (TPS) is based on onshore construction and testing of a total subsea system which consists of flowlines and control lines as well as facilities for manifolding and, if appropriate, for drilling and processing (see Fig. 8). The installation of such a system can be done by a pull or tow technique.

The TPS can be designed to meet any specific field requirement provided that the total system length and mass are within the practical limits to enable the pull or tow installation. The maximum single length pipelines/bundles which can be handled by using CDTM is approximately 7 km, and with bottom tow or off-bottom tow, the possible length is up to 20 km. The mass of an integrated manifold/towhead can be up to 500 metric tonnes or even more if required. To illustrate the flexibility in design of the TPS, and its adaptability to meet specific field development requirements, several concepts are described below (see Fig. 8).

Flowline bundle to single or cluster of wells

Usually several flowlines and an umbilical are required to connect a fixed production platform to a single satellite well. They can be combined in a bundle which is towed out using a tow method. Similarly a bundle configuration can be designed to envelop all flowlines to connect a multi-well template (or a cluster of wells) to the production facility. In this case all flowline connections are situated within the bundle towhead structure.

Flowline bundle with lateral entries

If the reservoir development requires 'spaced out' tree locations, a series of later entries can be provided along the length of the bundle. Each lateral entry will require a small manifold and control interface which can be retrofitted to the side of the bundle. This manifold will provide entry to an individual or to a bulk flowline and chemical injection as required. The control interface will connect with the control/data acquisition umbilicals.

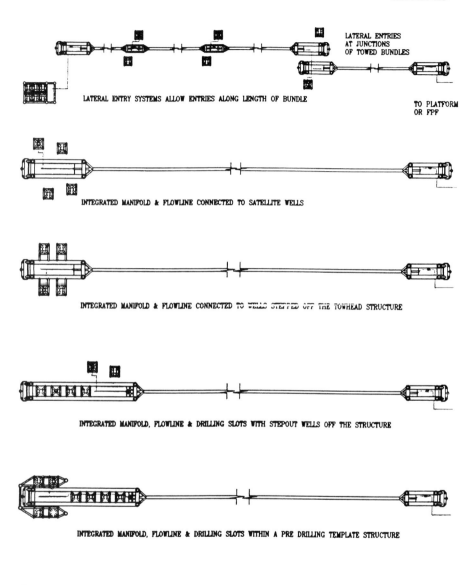

Fig. 8. Towed production systems

Integrated manifolding

For a (deep water) floating production system, the tie-in of a subsea satellite development based on the TPS concept typically consists of a flow and control lines bundle with an integrated production manifold at one end and an integrated riser base at the other end.

The bundle towhead structure is designed to accommodate the manifold to which the subsea trees can be tied in. Like a separate template manifold, this can provide, for example: dedicated or commingled flowlines system, full pigging loops, pig launch/receipt facilities etc., and controls and monitoring systems.

If the flowline bundle routing can also be worked out so as to pass an intermediate cluster of wells (between production facility and towhead/manifold), these can also be tied into the system by using a manifold structure integrated in the bundle at an intermediate location.

Detailed static and dynamic engineering analyses show that the characteristics of the bundle are not influenced significantly (during CDTM tow and installation) by the integration of a manifold at each end. The results show that manifolds with a dry weight of up to 500 tonnes in combination with a bundle length of 7 km and a bundle diameter of up to 40" can be transported and installed in the field using existing CDTM techniques.

Integrated drilling

Drilling slots can also be integrated in a towhead structure. Where permanent guide bases are utilised, all subsea hook-ups and tie-ins (from base of platform through manifold to trees) may be avoided, thereby resulting in an integrated remote drilling and manifolding system which is totally pre-connected and tested onshore with the required control and data systems and flowlines.

Integrated processing

A towhead assembly or intermediate structure can likewise be designed to accommodate processing facilities. These can be subsea pumping (booster station) or separation systems.

Integrated risers

Integrating risers with the TPS raises the possibility of stand-alone production systems, not necessarily tied back to an existing infrastructure. Used in conjunction with a floating production facility that incorporates processing plant, and in conjunction with export to a tanker via an export riser, the TPS is a rapid deployment solution for deep water or marginal field developments.

Technical advantages

The development of Pipeline Bundles and Towed Production Systems offers important technical and economic benefits, which makes the concept ideally suitable for satellite developments both in shallow and very deep water. When comparing the CDTM technique with reel or laybarge/vessel installation techniques, the following advantages are apparent.

Pipeline bundle

The carrier pipe, together with two end bulkheads, provides a closed container which, when filled with inhibited seawater after installation, gives a corrosion-free environment. This reduces or avoids the need for individual corrosion coating of the flowlines.

The submerged weight of the flooded bundle is sufficient to provide on-bottom stability and resistance to upheaval buckling without trenching or burial.

The carrier pipe provides protection against external hazards such as trawlboard impacts.

Thermal installation of the flowlines, when required, can easily be accommodated in the carrier design. Various insulation systems are available dependent on circumstance, such as neoprene, polyurethane elastomer, or filling the annulus with an insulating gel or cement slurry.

Exotic material such as Duplex stainless steels or composites can be used with a high degree or reliability due to the optimum fabrication conditions and quality control facilities inherent in onshore fabrication.

Onshore fabrication offers further benefits, such as working in a climate-controlled environment, rigorous inspection, and full onshore testing and commissioning which avoids expensive offshore hold-ups. All of these together provide a high level of confidence.

Further benefits may include a far less complicated umbilical design by laying the individual hoses and cables in the carrier without armouring.

Bundles reduce the pipeline corridor width to a minimum and avoid unnecessary congestion, allowing for future developments and minimising the risk of anchor patterns.

Integrity monitoring of the bundle can be provided by measuring the annulus pressure.

The carrier pipe provides an initial barrier against any leaking product and thus minimising the risk of environmental pollution.

Present carrier designs can be made suitable for deep waters (>1000m).

Towed production system

When the (template) manifold and/or riser base is integrated into the bundle this will offer further benefits such as

    (a) free transportation and installation of the manifolds, since they are an integral part of the bundle;
    (b) full testing and commissioning onshore of the total system, from the riser base on one end to the manifold on the other end, giving a maximum level of confidence and avoiding expensive delays offshore;
    (c) the reduction of the number of subsea tie-ins and spool installations, thus minimising subsea work and increasing reliability;
    (d) easy accommodation of USV systems and controls; the manifold structure can be designed to provide permanent in situ protection and it can be used to support any additional protection structures required to protect the jumpers to satellite wells or future xmas trees;
    (e) the concept allows for maximum flexibility of the development, both in location and in time, minimising capital expenditure.

Economic advantages

The economics of Pipeline Bundles and Towed Production Systems (TPS) are dependent on the particular field to be developed. However, from various cost analysis carried out to date there are potential savings to be achieved with a Towed Production System approach when compared with conventional field developments.

Procurement
Controls system can be simplified. The number of connections is reduced. Cheaper insulation systems can be used.

Fabrication
The structure for the Towed Production System facility (compared to one designed to be lifted) weighs less and the protection is integrated into the towhead structure.
    Welding of exotic material costs less onshore. Internal coating systems can be economically used where required.

Installation
The manifold and associated production elements get installed free with the bundle. No heavy lift vessels are required.
    The flowlines do not need to be trenched.
    Supplementary stabilisation (rock dumping) is not required for protection or to restrain any upheaval buckling tendency.

No hook up between the individual elements of the system.
Offshore spread comprises fewer and less costly vessel for a shorter period of time.
Disruption of other vessels in the field is reduced.
Reduced seabed congestion around offshore facilities.

Testing and commissioning
The high degree of onshore integration and testing leads to considerable savings offshore as well as giving a higher confidence in the system during commissioning.

Operating expenses
Reduced number of lines for external annual inspection.
Improved monitoring of the flowlines during operation from annulus monitoring.
Thermal efficiencies can be improved after commencement of operations by introducing gels into the annulus at later dates.
Easier decommissioning and abandonment.

Further applications
In August 1992 a study was completed that proved the feasibility of applying CDTM in very deep waters. The PETROBRAS 'MARLIM' complex anticipates flowline installation in water depth of 1000m and the 'ALBACORA' field in water up to 1400m water depth. These water depths require complex remote controlled systems which can only be constructed using remote controlled installation techniques. It is obvious that the highest possible systems reliability should be achieved. The concept of TPS, based on onshore construction and testing of a complete subsea system and reducing activities offshore as much as possible, is ideally suited for future developments in very deep waters.

Also in August 1992 the Phillips Embla bundle was installed between the Eldfisk and Embla platform. This was a fast track project only taking 8 months from contract award to installation. The Embla bundle is 5 km long and has a carrier of 24". The carrier contained a 14" duplex stainless steel gas pipe, which was a first application for duplex steel in bundles.

Presently ongoing is the Chevron Alba bundle, to be installed in May 1993. The Alba bundle has a carrier of 28" and a length of 2.7 km and it connect the Alba platform to the offloading tanker. The carrier contains an insulated 12" oil production line encased in a 16" sleeve pile and a 4" fuel line. One end integrates the pipeline end manifold (PLEM).

NORTH SEA INNOVATIONS AND ECONOMICS

Conclusions

The onshore fabrication and subsequent tow and installation of the towed production system, offer a reliable and economic method for development of satellite and marginal fields, as well as stand-alone production systems for both shallow water and diverless deep-water situations.

Towing pipeline bundles has proved to be a reliable installation method over the last 12 years. Recent experience has shown that substantial structures may be integrated with a bundle. The Piper/Saltire bundle installed in May 1992 weighed 5870 tonnes with a length of 6.7 km and structures on either side of 135 tonnes.

Onshore fabrication and testing provides increased reliability and substantially reduces offshore installation and commissioning costs.

The towed production system offers many technical advantages and proves to be 10 - 20% cheaper than traditional installation methods.

Discussion

A delegate wanted to know the limiting factors affecting the length of pipeline bundles for mid-depth tow. In response the speaker indicated that space at the launch site was the a constraint. The main constraint is that the longer the bundle, the harder it would be to control its position in the field because of currents.

A delegate added that if one thinks of reel barges then one should not be limited by the length of the bundle. Could bundles 6km long be laid, picked up again and joined together? The speaker agreed, provided the forces could be dealt with. Tow heads would be needed at both ends of each bundle, to connect individual lines. However, greater lengths have been achieved by joining two bundles together (Gannet & Osprey) but the connection is more effectively completed on the seabed.

A speaker commented that there were many such possibilities. Three lengths were laid for Frigg MCPO1 by a lay barge then pulled round by bottom tow. A length of pipeline laid for Hamilton Argyll had been similarly relocated.

A delegate then enquired about diverless tie-in of pipelines. The speaker stated that the Elf East Frigg bundle had been tied in in the off-bottom mode after installation by Controlled Depth Tow, without the use of divers.

A delegate wanted to know whether or not a tug had been used to launch a bundle and if so whether the tug was anchored or free riding. He also asked whether the chains with flexible wrappings gave uniform drag and lift during Controlled Depth Tow. The speaker replied that a tug had indeed been used for Phillips Embla and that it was anchored, even though it had power available should it have needed it. Flexible sheaths were adapted after model tests and performed well.

Ensuring the integrity of ageing offshore pipelines using on-line inspection tools and fitness-for-purpose methods

J. C. BRAITHWAITE and P. HOPKINS, British Gas plc, UK

Introduction

Transmission pipelines are very safe, with fatalities per ton kilometre transported much lower than any other means of transportation (ref. 1).

However, failures do occur (refs 1-5), and these failures are usually caused by defects in the pipe, or in-service damage which results in defects. In recent years there has been a growing awareness amongst pipeline operators that the integrity and maintenance of their strategic pipeline assets is paramount both for environmental and financial reasons.

An offshore pipeline which has been designed, constructed and tested in accordance with an adequate construction code and appropriate component manufacturing standards, should be free from significant defect at the time of commissioning.

To ensure its continued fitness-for-purpose requires

(a) that the pipeline continues to be operated within its design envelope

(b) that it is protected from any time-dependent degradation processes; these are primarily corrosion, external interference and ground movement.

The effectiveness of on-line inspection systems which can indicate the rate of deterioration of a pipeline goes beyond inspection. Such a method makes it possible to control deterioration by making adjustments to the CP system (onshore) in affected areas at an early stage, the inhibitor programme in offshore lines, and repairing coating damage where it has occurred. The life of a pipeline may thus be prolonged considerably over its original design life.

Accurate information on pipeline defects also permits continuous assessment of the safe operating parameters of the pipeline by means of fitness-for-purpose studies using modern fracture mechanics techniques (ref.6).

Since 1974 British Gas has made a major commitment to the study of the

Fig. 1. 813 mm magnetic inspection vehicle

causes of pipeline failure and to the development of inspection techniques to quantify these defects. These systems have operated since 1979 initially in its own onshore transmission system, then in its own offshore pipelines, and since then British Gas has inspected approximately 30 000 miles of pipeline world-wide, both on and offshore on a contract basis for a wide range of pipeline operators.

The main investment has been concerned with metal loss inspection using highly-developed magnetic flux leakage technology. This has led to a comprehensive range of metal loss inspection systems which will truly quantify metal loss in all parts of the pipeline and give the operator a clear and concise picture of the condition of his pipeline. Figure 1 shows a typical magnetic flux leakage tool.

This paper describes the magnetic flux leakage system used to detect metal loss defects, and also briefly mentions a new vehicle for assessing burial and weight coating loss. Finally the paper covers the methodologies available for assessing defects in pipelines, when they are detected by inspection.

Inspection

Evolution of the system

The magnetic systems were first upgraded to run offshore in 1982 and although they had high resolution at that time they have been continually upgraded in wall thickness capability operating range and data analysis capability to meet the demands of increasingly stringent inspection specifications.

The basic requirements of an on-line inspection system for the revalidation of pipelines can be formulated as follows. It must

(a) detect all significant defects,
(b) accurately locate the defects,
(c) determine the defect size,
(d) distinguish between true defects, spurious signals and natural pipeline features,
(e) avoid interference with pipeline operations.

While this list does not change with the type of pipeline and product, the characteristic form of defect does vary considerably from pipeline to pipeline according to its age, construction standard, product and product treatment. One of the lessons learnt in 10 years of offshore operation is that defect reporting systems suitable for one pipeline or pipeline operator are not necessarily suitable for another.

Another very significant evolutionary change is the involvement of the inspection contractor with the fitness-for-purpose assessment of the pipeline.

Originally it was considered that the inspection contractor could (and should) confine his role to reporting the location and physical nature of the defect and remain detached from the operational problem of sentencing the defect and the continued safe operation of the line.

However, an increasing number of pipeline operators request a full package extending through pipe preparation and cleaning inspection, defect reporting, defect sentencing and fitness for purpose report suitable for submission to a regulatory authority. The unique range of expertise available within British Gas allows this requirement to be met.

Magnetic tool enhancements

The basic principle of the magnetic flux leakage tool is well proven and has been extensively reported (refs 7 and 8). Its basic advantage is its extreme robustness when properly engineered and its ability to operate in both gas and liquid product pipelines and to tolerate significant quantities of debris, scale, wax, etc.

Although the original tool developed for British Gas onshore pipelines was produced to a high specification, being the first tool to utilise fully the capability of modern solid state digital electronics, it has proved capable of considerable performance enhancements under the pressures of commercial contract operation. Some of these are detailed below.

Range, data storage and power

These characteristics are closely linked. The initial large diameter (600mm plus) tools were designed to suit the 80 km range between launch and receive traps and took advantage of onboard data reduction techniques to achieve this range with a very fine inspection grid of approximately $1cm^2$. Approximately 400 megabytes of inspection data were recorded.

Current tools can inspect over 450 km in a single pass and the data storage capability has been enhanced to approximately 5 gigabytes. The inspection range was originally also limited by the capacity of the on-board battery packs. These are now supplemented when required by on-board power generation off the pipe walls, giving virtually indefinite range even at low product speeds.

Pipeline constructors are able to take advantage of these range enhancements, and offshore lines are being installed with over 800 km between pig traps in the confident expectation that they can still be inspected in one or two passes of an inspection tool. This has saved many millions of pounds in installing intermediate platforms and risers.

NORTH SEA INNOVATIONS AND ECONOMICS

Sensor design, welds and wall thickness variations
The basic requirement for accurate and consistent defect measurement means that a large number of sensor channels is required, and considerable trouble was taken to mount these on a suspension that remains in good contact with the pipe. This was originally intended to reduce the 'weld shadow' of uninspected pipe on either side of the girth weld to negligible proportions.

The system has been sufficiently successful to enable significant weld defects such as corrosion, lack of fusion and cracking to be detected.

Figure 2 shows a signal from a girth weld crack initiated during construction of an offshore pipeline.

The flux leakage system is primarily suited to the detection of discrete defects that disturb the magnetic flux permeating the pipe wall. Continuous areas of wall thinning only give a signal at the beginning and end of the section. Operation in pipelines with variable wall thickness, e.g. riser to sealine variations, lead us to add an auxiliary sensor system capable of measuring the wall thickness directly, and this system has been enhanced to give a direct measure of large areas of corrosion-induced wall thinning. This system was of great use when inspecting the Forties pipeline where inhibitor problems led to significant areas of wall thinning affecting complete pipe spools (refs 9-11).

Data handling and analysis
As mentioned earlier, the resolution and accuracy of the British Gas inspection vehicle means that a defect can be very precisely described in terms of depth, length and width, as well as location. Our early experience was in pipelines containing few defects, which led to a reporting system which very precisely described the location and size of each defect detected.

When British Gas began to inspect more severely corroded pipelines, a great deal of development work went into developing automatic defect discrimination and sizing techniques.

During the first stage of analysis, known as 'boxing', the software recognises each defect along the pipeline and ascribes to it a length, depth and width. Each combination of pipeline diameters and wall thickness has a sizing model which allows automatic prediction of defect size from signal collected. This sizing model is constructed from a combination of mathematical modelling and reference to a database 'library' of many thousands of real and manufactured defects. The defect listing which is subsequently produced acts as the base data for further examination.

The fracture mechanics implications of adjacent defects can be allowed for by specifying interaction rules to the boxed data in a process termed 'clustering'; the process allows the development of cluster plans which gather

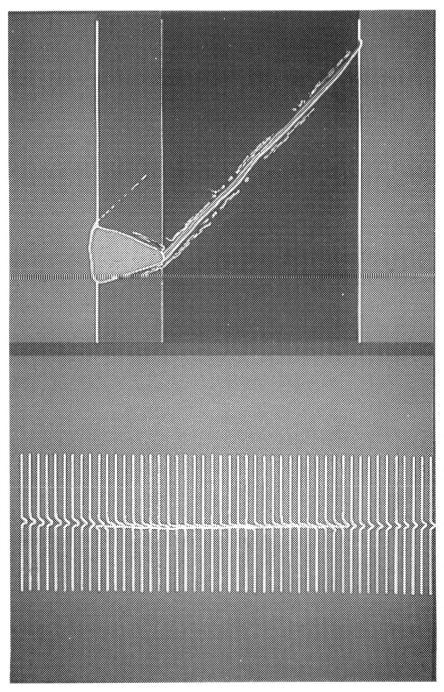

Fig. 2. Cracking indication from a girth weld

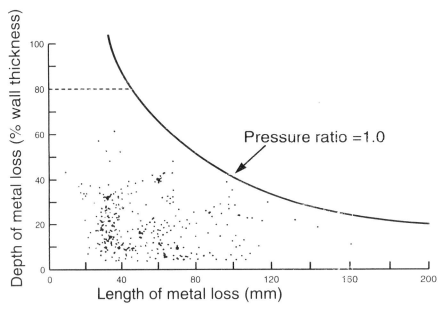

Fig. 3. Defect sentencing plot

Pressure Ratio	Grade	Upstream Weld	Absolute Distance (Metres)	Orientation (o'clock)
1.97	L★★	54680	215076.0	12-00
1.97	L★★	19020	73137.7	12-00
1.96	L★★	46760	183830.4	03-00
1.96	L★★	13520	51217.5	05-00
1.95	L★★	52240	205100.0	06-00
1.95	L★★	52120	204571.9	12-30
1.95	L★★	49860	195745.1	07-30
1.95	L★★	32280	125846.5	05-30
1.94	L★★	56130	221159.5	06-00
1.94	L★★	33740	132072.8	04-00
1.94	L★★	30050	117172.7	12-30
1.94	L★★	18750	72010.1	09-00
1.93	L★★	51870	203529.5	04-00
1.93	L★★	44440	174884.6	05-30
1.92	L★★	54600	214745.8	11-00

Fig. 4. Defect severity table

together all interacting defects before assessing their significance.

A further stage of analysis involves the calculation of failure pressure of each defect or group of defects. The calculation can be carried out automatically for each of many thousands of defects, using the surface flaw equation (see Eq. 1, later) if required, or relationships specified by the client, or those detailed in company or national repair codes.

The result is a series of reports (Figures 3 and 4) which tell the pipeline operator which of the defects in his pipeline are the most significant from a failure viewpoint. This is obviously a very powerful reporting tool as it allows the operator to prioritise his pipeline repairs.

This approach has been used by many of our clients with great success. One such client had identified 1809 defects in a 176 mile section of 34" pipeline, following the inspection of that pipeline using a first generation inspection tool.

Following inspection with the British Gas inspection vehicle the data were analysed using the techniques described above and a report was produced which gave the pressure sensitivity of all the defects in the line.

Using approved repair codes, the client was able to reduce the number of sites requiring repair to three.

Corrosion growth
With the increasing maturity of the inspection market, many pipelines are now being inspected for the second and third time. High resolution inspection enables signals measured from consecutive inspection to be compared and an accurate picture of corrosion growth to be presented.

Statistical techniques have been added to the discreet sizing analysis of individual defects, making use of the millions of readings recorded in each pass. This enables corrosion rates to be predicted with confidence from inspection intervals as short as 6 months (ref. 11).

Vehicle operation and performance
Since the inception of the project, British Gas has developed and built magnetic inspection tools for every pipeline diameter between 8" and 48".

We have inspected nearly 30,000 miles of pipeline world-wide. Of the many thousands of defects reported to clients, over 88% of those verified have fallen within our contractually guaranteed tolerances, the specification for location has been met in 97% of cases.

Burial and coating tool
As well as the usual onshore pipeline design requirements, offshore pipelines must have negative buoyancy to keep them in place on the sea bed. They

must be capable of withstanding impacts from ships' anchors and trawl boards and must contend with loss of support from the sea bed due to tidal current effects.

Offshore pipeline operators have adopted subsea surveillance methods to inspect for the following threats to pipeline integrity

(a) exposure of the pipeline on the sea bed
(b) damage to or loss of concrete weight coating
(c) the presence and nature of unsupported spans.

Current techniques employ such methods as side scan sonar, sub bottom profilers, ROV and diver visual surveys. These techniques, particularly diver and ROV surveys, are expensive to operate.

A pig-based system has obvious advantages. Firstly, the pig cannot drift unknowingly off the pipeline. Secondly, the quality and timing of the inspection is not affected by subsea visibility or weather conditions, and thirdly, shallow waters and intertidal areas can all be inspected in the same inspection mission.

British Gas has developed an inspection technique based on a neutron interrogation method to address this problem.

The core of the vehicle holds a neutron source, normally held within a radiation shield, but capable of being exposed when required. Once the source is exposed, neutrons pass through the pipeline steel and the concrete coating into the surrounding medium.

The neutrons interact with the surrounding material, producing radiation characteristic of the composition of that material. Some of the characteristic radiation travels back into the pipeline and is detected by sensing units mounted circumferentially around the pig. The data is then recorded by the on-board electronics.

Burial and coating vehicle
At the start of the development project a performance specification was set for the inspection system as follows: to identify

(a) complete burial of the pipeline
(b) progressive loss of cover with an angular resolution of +/- 20 degrees measured from the pipe centre-line
(c) the onset of loss of support of the pipeline
(d) loss of weight coating exceeding 10% in each 3D length of the line.

All these parameters are measured and reported at distance intervals in the range of 3 to 12 pipe diameters. The exact reporting interval selected depends on vehicle speed and the thickness of the weight coating on the pipe.

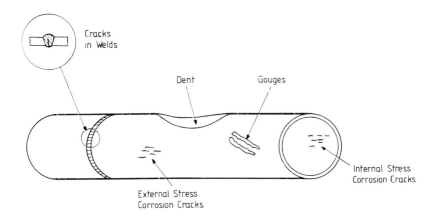
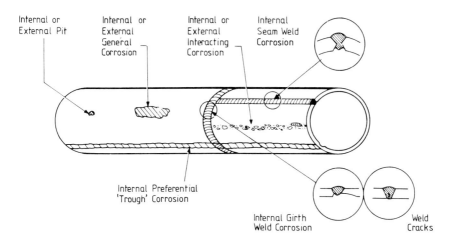

Fig. 5. Defects in transmission pipelines

British Gas had developed this tool initially in a 36" version for its Rough and Morecambe Bay offshore lines. Its performance has been reported in detail in ref. 12.

A 34" version has also been produced for a client with an oil pipeline and this inspection is ongoing.

Fitness-for-purpose

Inspection vehicles can accurately locate and size defects in a pipeline including mechanical damage, corrosion and girth weld cracks, Figure 5. The question the pipeline operator faces when presented with a defect problem or report is 'How significant is it?' Fortunately, there have been significant advances in fitness-for-purpose analysis methods and understanding in recent years. Now, most pipe body or weld defects in a pipeline can be assessed using these methods. These assessments can avoid unnecessary and expensive repairs or replacement.

This part of the paper presents the state-of-the-art in fitness-for-purpose methodologies for the assessment of defects expected or detected in offshore pipelines.

Defects in transmission pipelines and risers

Most defects or damage in pipelines and risers can be grouped into three categories (ignoring gross design/operational faults such as buckles or bend wrinkles).

Mechanical damage. It is generally accepted that mechanical damage ('third party interference') is the major cause of incidents in onshore and offshore pipelines carrying oil or gas (refs 5, 13 and 14), and such damage is often detected during periodic inspections.

Corrosion. As pipelines age, corrosion is increasingly becoming a problem. Pipeline operators throughout the world are facing multi-million dollar replacement or maintenance bills because of corrosion. For example, external corrosion in the Trans-Alaskan oil pipeline will cost $1.5 billion to repair (ref. 15) and the Forties oil pipeline in the North Sea has had to be replaced at a cost of $265 million because of internal corrosion caused by the presence of water in the oil (ref. 10). Additionally corrosion now causes most failures in the Gulf of Mexico (ref. 16), and reports of failures in Russia (ref. 17) and the UK (ref. 18) directly attributable to corrosion highlight a growing concern about their effect on the environment.

Therefore, mechanical damage may pose the greatest failure threat, but

Table 1. Simple acceptance levels for gouges in pipelines (<72% SMYS)

Gouge Length $2c/(Rt)^{0.5}$	Failure Depth, %t	Acceptable Depth (Safety Factor of 2 on depth), %t	Acceptable Depth (Using 100%SMYS as s in Eq.1), %t
1	77	39	37
3	49	25	14
5	42	21	11
7	40	20	10
>10	34	17	8

corrosion poses the greatest financial threat.

Weld defects. Two types of weld defect can cause problems in a pipeline: defects from the fabrication process, and defects caused or extended during operation. Fabrication defects in the longitudinal seam weld which survive the pre-service hydrotest will not generally effect the integrity of the pipeline, and consequently this weld does not require in-service inspection (refs 19 and 20). However this weld can be prone to preferential corrosion during service in some oil pipelines (ref. 11).

Pipeline girth weld defects can be detected during post-construction weld audits (ref. 21) or by on-line inspection using an 'intelligent pig' (ref. 22). Both fabrication defects (cracks) and in-service defects (preferential corrosion) at girth welds have been detected by these inspection pigs (refs 9 and 11).

The following section summarises the fitness-for-purpose methodologies for the assessment of all the above defects.

Methodologies for the assessment of pipeline defects

Mechanical damage. Mechanical damage to offshore pipelines can result in a dent, a gouge, or a combination of the two. This damage can be detected during surveys (e.g. by remotely operated vehicles) or detected and sized by on-line inspection (by calliper/intelligent pigs).

Pipeline steels are ductile, and pipebody defects do not generally pose a brittle fracture risk. Therefore, they can be assessed using simple limit load relationships. The significance of gouges can be assessed using the formula (refs 23-27)

$$s = s_f(1-d/t)/(1-(d/t) \times M^{-1}) \qquad (1)$$

where s is the pipeline hoop stress at failure, d is the defect depth, t is the pipewall thickness, R is the pipe radius, 2c is the defect length, and M is equal to $(1 + 0.4(2c/(Rt)^{0.5})^2)^{0.5}$.

For pipeline steels, the flow strength (s_f) is usually taken as between 110 and 115% of the yield strength of the pipe (refs 23 and 24). For steels limited to a yield to tensile strength ratio of 0.85 (ref. 6)

$$s \leq 1.09 \times \text{yield strength} \tag{2}$$

Simple acceptance levels for gouges can be obtained from Eq.1. For pipelines operating up to 72%SMYS, a gouge of any length whose depth does not exceed 34 to 40% (depending on the flow strength assumption) will not fail. From Eqs. 1 and 2 (ref. 6) Table 1 may be drawn up.

Equation (1) contains no explicit safety factors (ref. 6). Using the '100%SMYS' safety factor above is equivalent to only allowing defects that would survive a hydrostatic test to this level. Use of this large safety factor would ensure that the pipeline is in an equivalent condition to that following its pre-service hydrotest. A safety factor of 2 on depth is a simple engineering margin of safety to account for sizing errors, etc. Use of this safety factor is reasonable if a line is not to be rehydrotested or uprated.

Finally, three points should be emphasised when discussing gouges. The first relates to fatigue. Some pipelines are pressure-cycled, which can cause some defects to increase in size. In these lines, all part-wall defects should be assessed for fatigue, using standard methods (refs 20, 26 and 28). The second point relates to cracking detected in a gouge. A crack in the base of a gouge may be caused by severe fatigue loadings, or an associated dent. A cracked gouge should be treated with caution, and if superficial grinding does not remove the cracking, it should be repaired. Finally, gouges in pipelines operating at below 30%SMYS are unlikely to rupture (i.e. on failure they will leak and not extend along the pipe) (refs 24 and 26), unless they are in excess of a pipe radius long (highly unlikely). This 'leak before break' criterion is useful in determining the consequences of failure.

Plain smooth dents (with no associated loss of wall thickness defect) of depth up to 8% pipe diameter have little effect on the burst strength of linepipe (ref. 29). However, plain dents can exhibit fatigue lives below design requirements, and therefore should be assessed for fatigue failure (ref. 30). This is not a major restriction, as most oil and gas lines are not cycled over a large pressure range.

Dents containing seam welds can exhibit low burst strengths and fatigue lives. Dented seam welds should be repaired (ref. 29).

In summary, plain dents (away from seam welds) are not a problem in pipelines. Obviously, they should not be so big as to restrict flow or the passage of pigs, but all the research work on them indicates that they are insignificant.

The severest form of mechanical damage is a combination of a defect in a dent. The combined dent/defect can record very low burst strengths and fatigue lives (refs 29 and 30). The major problem with this type of defect is that it is a combination of a severe unstable stress concentration (a dent) and a severe stress raiser (a defect). Under pressurisation the dent decreases in depth, resulting in high bending stresses around and in the dent. This movement and high stress causes the defect to rapidly tear through the wall to produce a very low stress failure.

The failure stress of a combined dent/defect can be estimated using a semi-empirical relationship (refs 6 and 30)

$$s = s_f(2/Pi)(\cos^{-1}(\exp-(B^1(Y_1(B^2) + Y_2(B^3))^{-2})\exp(B^4)) \tag{3}$$

where
$Y_1 = 1.12 - 0.23d/t + 10.6(d/t)^2 - 21.7(d/t)^3 + 30.4(d/t)^4$
$Y_2 = 1.12 - 1.39d/t + 7.32(d/t)^2 - 13.1(d/t)^3 + 14.0(d/t)^4$
$B^1 = 1.5(Pi)E/(s_f^2 Ad)$
$B^3 = 10.2(D_o/2t)$
$D_o = (1.46D) - (0.0046R)$
$B^2 = 1-1.8(D_o/2R)$
$B^4 = (\ln(C_v)-1.9)/0.57$
$s_f = 1.15SMYS(1-(d/t))$

(E is Young's Modulus, C_v = 2/3 Charpy, D is dent depth, A is Charpy Fracture area.)

No safety factor is included in Eq.3. All angles are in radians, and the units are lbf/in^2, ft, lbf, in and in^2.

In summary, combined dents and defects pose the most severe threat to pipeline integrity. They can be assessed using a semi-empirical relationship, but care should be taken as they are unstable and prone to low burst strengths and fatigue lives.

Corrosion. Some intelligent on-line inspection pigs such as the British Gas pig can accurately detect and size corrosion in pipelines. The significance of pipe corrosion can be assessed using established analytical methods (refs 23-26) which have found approval in national codes (ref. 27). Equation (1) provides a conservative assessment criterion for corrosion of any length (ref. 31). Short length corrosion ('pits' of length up to 300% wall thickness) does not generally pose a failure threat to the pipeline, and any length of corrosion is insignificant at design stresses (H%SMYS), providing its depth does not

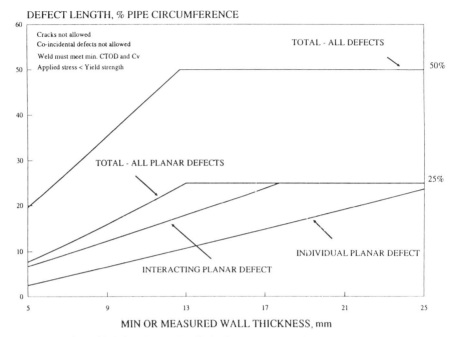

Fig. 6. Girth weld defect limits (wall thickness < 25 mm)

exceed the depths given in Table 1 above. The levels in this Table will be conservative for corrosion (refs 24 and 31), but provide reasonable 'screening' levels. Furthermore, corrosion of any length is unlikely to cause a rupture (i.e. it will only cause a leak) in pipelines operating at stresses below 40%SMYS (ref. 24). Irregular-shaped corrosion can be assessed by using the average depth in Eq.l (refs 23,24,31).

A general methodology for the assessment of corrosion detected during on-line inspections is given in ref. 11. This methodology, in conjunction with a fitness-for-purpose analysis, can allow corroded pipelines to continue to operate safely, and can also be used to plan inspection levels and intervals for intelligent pigs.

In summary, fitness-for-purpose assessments of pipeline corrosion are now routine. The assessments can use sidelines in codes or alternative methods and can easily be incorporated into an on-line inspection programme. Operators who have corrosion reported following an on-line inspection should be reassured by the fact that their pipeline is very tolerant to corrosion, and its significance can be quantified.

Girth weld defects. Pipeline girth welds have a good operating record, and are not a major cause of pipeline failure (ref. 3). They are fabricated to stringent standards (e.g. API 1104, BSI 4515, CSA Z184) which may account for their good safety record. However, the defect acceptance levels in these standards vary significantly even when they are based on fitness-for-purpose methods (ref. 6).

Reference 6 gives full details of the assessment of girth weld defects and concludes that even grossly defective girth welds will fail at high stresses predicted by plastic collapse formulae (e.g. ref. 25). These equations can be used to develop limits for girth weld defects. Figure 6 (ref. 6) provides general girth weld defect limits.

Some examples of the application of fitness-for-purpose to offshore pipelines.
Fitness-for-purpose methods have been applied to defects in pipeline constructions on many occasions (ref. 21): at the design stage (e.g. for setting girth weld limits (refs 19 and 20) or pipeline toughness requirements (ref. 34)), during post-construction audits (refs 20 and 21), for pipeline uprating (ref. 35) and following in-service inspection of welds (ref. 22) and the pipebody (ref. 11).

Girth weld defects in river crossings, offshore oil and gas lines and in-field pipelines and risers have been assessed using fatigue and fracture mechanics methods (ref. 20), as outlined above.

Seam weld defects have been assessed using the above methodologies (refs 20, 22 and 35), and crack propagation criteria have been used to set toughness requirements in offshore pipelines (ref. 34).

The methods are now regularly used to assess the significance of mechanical damage (refs 35 and 36) and corrosion (refs 12 and 35) detected in a pipeline following an in-service inspection by an intelligent pig.

It should be emphasised that the quality, accuracy and reliability of the assessment depends on the quality, accuracy and reliability of the input data, the engineer conducting the assessment, and the methodology adopted. This part of the paper has concentrated on the methodologies. It is the responsibility of the pipeline operator to acquire a competent engineer and provide all the necessary data. The three elements of an assessment (input data, engineer, methodology) are of equal importance.

Concluding comments

This paper has presented the background to, and performance of the British Gas vehicles to measure metal loss, and to measure burial and weight coating loss. It has also described the fitness-for-purpose methodologies available for the assessment of defects in offshore pipelines. Most pipeline defects can be reliably assessed following their detection.

The fine safety record of pipelines is partly due to their high tolerance to defects. Large dents, gouges, corrosion and cracks can be tolerated by a pipeline, and operators should not embark on an expensive (and sometimes dangerous) repair or replacement policy of a defective/damaged pipeline without first checking the defect/damage significance using the above methodologies.

The combination of on-line inspection technology, and fitness-for-purpose methodologies can ensure the continuing safe and efficient operation of ageing offshore pipelines. It is worth ending the paper by repeating the words of the Head of the UK Pipeline Inspectorate (ref. 5): 'One of the major issues for the (offshore pipeline) industry will centre around the extension of the operating life of many of the older pipeline systems... internal inspection by intelligent pigging will be the key to demonstrating the continued fitness-for-purpose of these systems.

Acknowledgements

The authors would like to thank British Gas for permission to publish this paper, and all their colleagues in the Research and Technology Division who have contributed to this paper.

References

1. ANDERSON T. and MISUND A. Pipeline Reliability - An Investigation of Pipeline Characteristics and Analysis of Pipeline Failure Rates for Submarine and Cross Country Pipelines. *Journal of Petroleum Technology*, April 1983.
2. CANNON A. G., LEWIS R. C. and SCRIVENER C. The Reliability of Pipe Systems Operating in the British Sector of the North Sea. *Reliability '85 Conference*, England, Paper 4A/R, July 1985.
3. EIBER R. J., JONES D. J. and KRAMER G. S. Analysis of DoT-OPSR Data from 20 Day Incident Reports, 1970-84. *7th Symposium on Linepipe Research*, AGA, Texas, Oct. 1986, Paper 2.
4. STRATING J. A Survey of Pipelines in the North Sea Incidents During Installation, Testing and Operations. *13th Offshore Technology Conference*, Texas, May 1981, pp25-32.
5. ADAMS A. The UK Experience in Offshore Pipeline Operations. *Pipes and Pipelines International*, March-April 1992, pp9-14.
6. HOPKINS P. The Application of Fitness-for-purpose Methods to Defects Detected in Offshore Transmission Pipelines. *Conf. on Welding and Weld*

Performance in the Process Industry, London, April 1992.
7. BRAITHWAITE J. C. Operational Pipeline Inspection. *IChemE Symposium (Series 93) on Assessment and Control of Major Hazards*, April 1985.
8. JACKSON L. and WILKINS R. *The Development and Exploitation of British Gas Pipelines Inspection Technology*. Inst. Gas Engineers, Communication 1409, IGE Autumn Meeting, London 1989.
9. SOWERBY T. In-Line Inspection Prompts Forties Line Replacement. *Oil & Gas Journal*, 17 June 1991, pp27-40.
10. LONDON C. J. The Forties Export Pipeline Project. *Pipes & Pipelines International*, May-June 1991, pp7-13.
11. HOPKINS P. The Assessment of Pipeline Defects Detected During Pigging Operations. *Conf. on Pipeline Pigging and Integrity Monitoring*, Aberdeen, Nov. 1990.
12. HOLDEN E. M., PAIGE D. M. and BRAITHWAITE J. C. On-line Inspection of Offshore Pipelines for Burial and Loss of Weight Coating. *Proc. of 18th World Gas Conference*, IGU, Berlin, 1991, paper IGU/C7-91.
13. Gas Pipeline Incidents. *Pipes and Pipelines International*, July-August 1988, pp11-14.
14. HOPKINS P. Interpretation of Metal Loss as Repair or Replacement During Pipeline Refurbishment. *Proc. of European Pipeline Rehabilitation Conf.*, London, May 1990, Paper 8.
15. KEEN J. Corrosion Forces Repairs to Oil Pipeline. *USA Today*, 5 Feb. 1990.
16. MANDKE J. S. Corrosion Causes Most Pipeline Failures in Gulf of Mexico. *Oil & Gas Journal*, 29 Oct 1990, p40.
17. PEEL Q. Soviet Oil and Gas Accidents Add to Growing Power Crisis. *Financial Times*, 17 July 1990.
18. Bromborough Oil Line Leak Report Published. *Pipes & Pipelines International*, Jan-Feb 1991, p3.
19. JONES D. G., HOPKINS P. and CLYNE A. J. Assessment of Weld Defects in Offshore Pipelines. *Conf. on Offshore Pipeline Technology*, Stavanger, Norway, Jan. 1988.
20. JONES D. G. and HOPKINS P. Methodologies for the Assessment of Defects in Offshore Pipelines and Risers. *10th Int. Conf. on Offshore Mechanics and Arctic Engineering*, Stavanger, Norway, June 1991.
21. KNOTT J. F. and HARRISON J. D. Fundamentals of Fracture in Pipelines. *Proc. Fracture in Gas Pipelines*, Moscow, March 1984, p1-26.
22. NESPECA G. A. and HVEDING K. B. Intelligent Pigging of the Ekofisk-Embden 36in Gas Pipeline. *63rd Annual Tech Conf. and Exhibition of the Society of Petroleum Engineers*, Houston, Oct. 2-5 1988.
23. KIEFNER J. F. *et al. Failure Stress Levels of Flaws in Pressurised Cylinders.*

ASTM STP 536, 1973, pp461-481.
24. SHANNON R. W. E. The Failure Behaviour of Linepipe Defects. *Int. J. Press. Vess. & Piping*, (2), 1974, pp243-255.
25. MILLER A. G. Review of Limit Loads of Structures Containing Defects. *Int. J. Press. Vess & Piping*, (32), Nos 1-4, 1988, p195.
26. FEARNEHOUGH G. D. and JONES D. G. *An Approach to Defect Tolerance in Pipelines. Conf. on Tolerance of Flaws in Pressurised Components*, IMechE, London, Paper C97/78, 1918.
27. *Manual for Determining the Residual Strength of Corroded Pipelines.* ANSI/ASME B.31 G-1984, ASME 1984.
28. *Guidance on Methods for the Derivation of Defect Acceptance Levels in Fusion Welds.* BSI PD 6493, BSI, 1991.
29. HOPKINS P., JONES D. G. and CLYNE A.J. The Significance of Dents and Defects in Transmission Pipelines. *Conf. on Pipework Engineering and Operation*. IMechE, London, 1989, Paper C376/049.
30. HOPKINS P., CORDER I. and CORBIN P. The Resistance of Transmission Pipelines to Mechanical Damage. *Conf. on Pipeline Reliability*, CANMET, Calgary, Canada, June 1992.
31. HOPKINS P. and JONES D. G. A Study of the Behaviour of Long and Complex-Shaped Corrosion in Transmission Pipelines. *Offshore Mechanics and Arctic Engineering Conference*, OMAE-92, Calgary, Canada, June 1992.
32. MCDONALD K. and HOPKINS P. Review of Literature Relevant to Acceptability of Non-Planar Imperfections in Pipeline Girth Welds. To be published, 1993.
33. HOPKINS P., PISTONE V. and CLYNE A. J. A Study of the behaviour of Defects in Pipeline Girth Welds: The Work of the European Pipeline Research Group. *Conf. on Pipeline Reliability*. Calgary, Canada, June, 1992.
34. JONES D. G. and NOKLEBYE A. Zeepipe Fracture Toughness Requirements. *Pipeline Technology Conference*, Ostende, Belgium, 1990, pp10.17-10.23.
35. JONES D. G. and NESPECA G. Uprating an In-service Offshore Pipeline. *ASCE Int. Conf. on Pipeline Design and Installation*. March 1990, Las Vegas.
36. BROWN P. J. 10 Years of Intelligent Pigging: An Operator's View. *Pipeline, Pigging and Inspection Technology Conf.*, Houston, Feb. 1990.

Discussion

A speaker asked what the main reason for repairs on pipelines would be. In his reply the speaker stated that for offshore pipelines it would be internal corrosion whereas it would be external corrosion and interference damage for onshore pipelines.

A delegate asked about the speed of the pig and whether or not the pigs were susceptible to damage. The speaker explained that the pig is normally run at the product speed (whether it is oil or gas). The maximum pig speed is 5m/s. In heavily loaded gas lines (e.g. Norwegian North Sea lines where the gas speed can be up to 30m/s) then the gas needs to be slowed to the pig speed. The minimum velocity would be ½ to ⅔ m/s, with the majority of oil product flow rates being in the range ½-2m/s. Responding to the second part of the question the speaker stated that well designed inspection tools run at high transmission pressures do not experience much structural damage, although rubbing components experience wear and tear. Pigs run at low pressures - e.g. cleaning pigs run during pipeline commissioning - and may be vulnerable to damage particularly if the back pressure is not well controlled. Any damage is more likely to occur in the older pipelines which have not been designed for pigging. Pipelines up to 1,000 km can be inspected by pigs, for example the 800 km Zeepipe line.

The speaker was then asked if corrosion allowances could be reduced now that good data were available on the condition of pipelines. The speaker replied that one recent pipeline has no corrosion allowance because it will be regularly inspected by intelligent pigs.

A delegate wanted to know the minimum size that can be inspected by use of a pig. The speaker stated that at present it is an 8" pipe but there is currently a pig being developed for a 6" pipe.

A delegate then asked the speaker to comment on the use of the outlined technique in comparison to the use of an ROV. The speaker explained that the comparison could only be in terms of pig versus ROV side scan sonar survey of the weight coat loss from the pipe. The pig-based technique gives more accurate and detailed information in more difficult conditions than the ROV. He added that the pig system does tend to be more expensive for short pipelines.

Engineering aspects of jacket toppling as a means of platform abandonment

S. WALKER and J. WILLIAMS, SLP Engineering, UK

Introduction

A number of different approaches have been considered for the deconstruction and subsequent disposal of steel jacket structures. The usual options include partial or complete removal, either piece-meal or in large sections, or toppling. The partial removal and toppling alternatives may be considered for platforms in deeper waters, whereas complete removal is the only option for decommissioning of platforms in shallow water.

There is now a limited but nonetheless growing body of experience of decommissioning platforms in deeper waters, initially in the Gulf of Mexico and more recently in the North Sea. Although many of the engineering aspects are still in the development stage, valuable lessons have been learned from recent projects.

In this paper, a number of the key aspects of the deconstruction of platforms in deeper waters are examined, and solutions to some of the major problem areas are put forward. The paper examines both practical design considerations, and the advanced numerical modelling which must be performed. These techniques are illustrated by reference to a design example.

Design criteria

A primary objective in the deconstruction of a platform is that the statutory requirement for free water above any remaining debris should be achieved. The current consensus of opinion favours dividing the North Sea into two zones by a line drawn approximately east by north-east from St. Fergus Head. This line corresponds roughly to the 100 metre isobath. Above this line, 75 metres of clear water is required, whilst below the value is 55 metres (see Figure 9).

It is of vital importance that the deconstruction operation should be reliable, because it would be very expensive and probably also hazardous to deal with a partially failed structure. There is also little opportunity to take corrective action during a deconstruction operation, if events do not proceed as planned. To fulfil these requirements, the following criteria should be adopted.

(a) Fail safe systems should be used where possible, so that the failure of any component will not result in the primary objective not being achieved.
(b) Critical systems should be duplicated.
(c) Untried technology should be avoided, unless absolutely necessary.
(d) It will not be possible to engineer all aspects of the work in full detail and, therefore, the engineering work should be concentrated on those aspects judged to be most critical.

The appropriate design philosophy entails a mixture of conventional engineering principles, such as might be employed in the design of permanent structures, and the more innovative approaches, which are to be found in some demolition projects.

Engineering aspects of toppling

The decommissioning of steel platforms involves deconstruction followed by removal to a disposal site. The key question in planning the deconstruction is to decide upon the optimum number and size of sections to be removed - this is a trade-off between the amount of underwater cutting to be performed and the size of the lifting tackle required (either in the form of lift barges or buoyancy tanks).

An option which is particularly attractive, for locations where it is permitted, is jacket toppling. The first main advantage of this is that the deconstruction and disposal are combined into a single operation. Secondly, toppling is effectively a 'one piece' removal, but without the need for heavy lifting equipment.

The structural engineering is carried out so that, by a combination of severing members, forming hinges and by applying destabilising forces, a toppling mechanism is formed which leads to the controlled failure of the structure. The design criteria for this are as follows.

(a) The mechanism should be arranged so that, even if unexpected circumstances arise, the platform will come to rest on the seabed leaving the desired clearance above the structure.
(b) The formation of unpredictable failure mechanisms should be precluded at all costs.
(c) Underwater working should, to the extent practicable, be minimised.
(d) It is vital that parts of the structure which are required to remain intact during the toppling mechanism should not be damaged, for example by

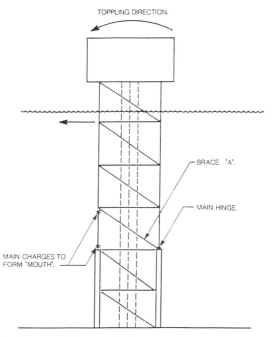

Fig. 1. Riser platform toppling mechanism

a temporary overload or by the forces from explosive charges.

Several variants of platform toppling have been studied (ref. 1) of which a few have been put into effect on actual deconstruction projects. In one project, buoyancy tanks were used in the controlled toppling of a steel jacket and the remains of the platform were used to make an artificial reef (ref. 2). Whilst this may have particular attractions for certain situations, it is not the preferred solution for a straightforward toppling operation since the method does not comply with the 'fail safe' criterion (e.g. the loss of a buoyancy tank would lead to an uncontrolled failure mechanism) and there is considerable underwater working involved in the attachment, and subsequent removal, of the buoyancy tanks.

A more straightforward toppling operation, undertaken in the Gulf of Mexico (ref. 3) in 1987, involved severing the piles at the mudline and pulling the structure to failure using cables anchored to the seabed. The structure was an 8 pile jacket standing in 350 feet (75m) water depth, weighing a total of 3500 tons. The operation was executed according to plan, but it is interesting to note that the actual pulling force of 290 tons was 50% higher than predicted.

Whilst this latter scheme could be applied to North Sea structures, several modifications would be desirable. Firstly, many of the larger North Sea structures have a considerable number of piles (typically in excess of 20) and so it is better to cut the legs above the piles. Secondly, it is advantageous to arrange the failure mechanism so that the self weight of the structure contributes to the destabilising moment, both to reduce the required applied forces from cables and to create a more reliable mechanism. These features are realised in the toppling mechanism illustrated in Figure 1.

Toppling of topsides with the jacket structure

Two possible scenarios are available with the toppling option: overturning of the jacket with the topsides still intact or removal of the topsides prior to jacket toppling.

Toppling of the jacket and topside together obviously offers the advantage of a one-stage deconstruction operation. In addition, the topside raises the height of the centre of gravity of the structure and thus decreases the stability of the platform during overturning. However, there are some drawbacks with the combined deconstruction scenario.

(a) The forces on the topside as it hits the water may be high and could break off items such as flare booms, drilling derricks and communications aerials - consideration should be given to removing these items before toppling.

(b) All loose items must be tied down to prevent them breaking free.

(c) All process equipment must be clean and free from toxic and dangerous materials. This is particularly important with piping and large vessels which will need to be punctured to reduce restoring (buoyancy) forces when the topside reaches sea level. An attractive option is to refill the vessels with sea water before toppling to increase the overturning moment and eliminate buoyancy forces. Checks will, however, need to be made to quantify the effect that this has on hinge forces.

The Joint Industry Project on Platform Abandonment (ref. 1) addresses the third item in detail and covers the required degree of cleaning, purging and removal of equipment under the categories

(a) hazardous systems - hydrocarbon handling and processing
(b) non-hazardous systems - utilities
(c) toxic and other hazardous chemical systems
(d) electrical systems.

To illustrate the extensive activities involved in fully decommissioning topsides, the eight operations required for hazardous systems were identified as

NORTH SEA INNOVATIONS AND ECONOMICS

(a) depressurise and purge
(b) drain
(c) water flush
(d) second drain
(e) cleaning (water jetting/steam cleaning)
(f) water fill
(g) inerting (nitrogen purge)
(h) isolation.

Most cleaning and purging is likely to be carried out in-situ if combined topsides/jacket toppling is adopted. However, with other deconstruction options for the topsides, although an initial 'first clean' of all systems would be carried out on the platform, this could be followed by onshore operations or use of an adjacent barge for secondary cleaning and scrap processing.

Underwater cutting techniques

In a platform deconstruction there is a substantial amount of cutting of members to be performed underwater, and the use of explosives for this purpose is becoming an established technique. Some of the relative advantages and disadvantages of this method, viewed from the structural engineering standpoint, are discussed below.

A major advantage of the use of explosives for cutting is the reduced amount of underwater working, compared to mechanical or thermal cutting systems. Additionally, after the explosives have been placed and fused, all of the required cuts may be made simultaneously. With conventional systems, the members are cut one by one, with the cutting operation possibly continuing over an extended period. The structure is thus in a weakened condition for a longer period, and there is an increased possibility of storm damage occurring.

On the debit side, because the cuts occur simultaneously, there is no chance to take corrective action should the operation not proceed according to plan. One of the major potential risks is that a critical charge may not detonate, leading to an incomplete collapse of the structure. It is prudent to duplicate critical charges and their associated systems to counter this possibility.

The major problem associated with the explosive cutting is the potential for unwanted damage caused by the explosion, both to the environment and to parts of the structure which are required to remain intact during the toppling. Steps must therefore be taken to minimise this damage.

(a) Undue conservatism should not be used in sizing the charges.
(b) Shaped charges should be used where possible for cutting steel mem-

bers, since they are more efficient than bulk charges.
(c) The charges should be placed as far away from critical members as possible.
(d) Consideration should be given to placing charges inside members - this gives the most effective use of the explosive, and partly contains the explosive forces.
(e) Measures should be taken to protect critical members, for example by internal grouting or by flooding the members concerned.
(f) Appropriate numerical simulations should be used during the engineering phase to investigate the effects of the explosive forces.

By using these techniques, it is possible to eliminate the uncertainty associated with explosive cutting, and to engineer reliable schemes for the deconstruction of steel platforms.

Design study - toppling of steel platform

Introduction

The considerations discussed in this paper will be illustrated with reference to a design study on the demolition of a steel riser platform. The platform forming the subject of this study is a 4 leg structure standing in deep water. This investigation considers toppling with the topsides in place. The wells and the conductors were assumed to be grout-filled and to be demolished with the platform.

Selection of the mechanism

To reduce the amount of steel to be cut in severing the legs, it is possible for the legs to be cut just above the pile sleeves. The positioning of the cuts leaves the base of the structure intact. The most effective means of forming a mechanism is to remove a length from two adjacent legs, thereby forming a 'mouth', and letting the self weight of the structure act as a destabilising force.

Considerable interest centres on the braces at the level of the mouth (see member 'A' in Figure 1). Removal of these braces by explosive charges could lead to uncertainty in the position of the hinges in the two remaining legs, and to an unpredictable failure mechanism. By leaving the brace members in place, the hinges will be constrained to form in the legs just above the pile guides. To assist in the formation of these hinges, light charges were considered at the hinge locations with the objective of weakening the joint cans.

The lower members above and below the hinge would be flooded in order to resist radial pressure loads and to increase the overturning moment.

In studying the mechanics of the toppling, it was apparent that there will be relative movement between the conductors and their guide frames, and

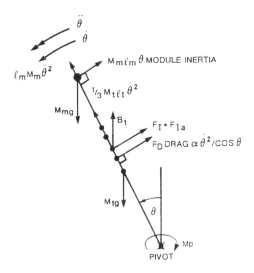

Fig. 2. Forces acting during toppling

that this will have to be accounted for.

Behaviour of the conductors

The behaviour of the conductors is deemed to have a potentially major influence on the toppling behaviour of the structure, and this was therefore studied in some detail. The conductors are 30" in diameter and filled with grout, which represents a very significant additional strengthening of the platform. It was calculated that the conductors could prevent the toppling from taking place and that it would not be possible to sever the conductors with reasonable quantities of explosives.

There are two basic mechanisms by which the conductors could prevent toppling, namely bending resistance and the development of axial forces.

Calculations indicated that the bending strength of the conductors is not a problem, since the effective lever arm is adequate to ensure the formation of plastic hinges. Regarding the possible axial load, it was computed that each conductor could support a force of approximately 600 tonnes between adjacent conductor guide frames. This would be sufficient to halt the platform toppling.

This axial force could develop if the conductors were prevented from sliding freely through the guide frames. It is also possible that the conductor connectors could catch on the guide cones.

A more detailed study revealed that if the conductors caught on the guide frames, the latter would fail without halting the platform toppling. However, if the guide frames fail, there is nothing to pull the conductors over with the platform. It is therefore proposed that wire ropes should be placed around the platform at two levels, to confine the conductors and ensure that they fail as intended.

Structural analysis

A range of detailed analyses were performed to investigate the major aspects of the toppling. The principal analyses were

(a) determination of the rigid body motions
(b) structural analysis of the platform at various stages during the toppling
(c) analyses to determine the explosive loadings on the structure of the platform and the local response.

The analysis of rigid body motions utilised a single degree of freedom model, in which the principal unknown is the angle of rotation of the structure (refer to Figure 2). The objectives of this analysis were to calculate time history plots, in order to generate inertial loads, drag forces and centripetal accelerations for inclusion in the main structural analyses. This analysis also produced time histories of the pivot forces, to enable the integrity of the hinges to be confirmed. The model accounts for the principal features of the system, ensuring that the results can be used with some confidence, and yet is sufficiently simple to enable parametric studies to be performed with a minimum of computing effort. As an example of the outputs of the analysis, time histories of the angle of rotation and the reactions at the pivots for a typical set of governing conditions are shown in Figures 3 and 4. It may be seen that the toppling takes some 20 seconds to achieve a 60% angle of tilt (at this stage the toppling may be regarded as virtually complete, since the platform would inevitably proceed to ultimate failure).

The main structural analyses were performed for various 'snapshots' in time to determine the member forces and to confirm that critical members will not fail prematurely. In theory, these analyses should be performed as transient, dynamic analyses; however, in practice, sufficient accuracy is obtained from quasi-static analyses in which a self equilibrating set of forces is applied to the structure. Several of these forces (e.g. inertia and drag) are dependent on the platform motions, which were obtained from the 'rigid body' analysis described above. The structural analysis clearly demonstrated the major differences in the loadpaths through the structure during the toppling, but confirmed the adequacy of the basic structure to resist these loads.

NORTH SEA INNOVATIONS AND ECONOMICS

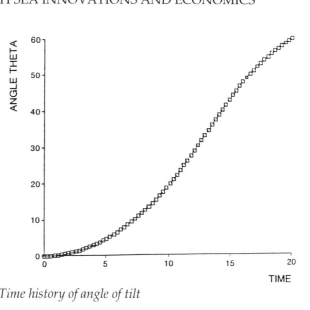

Fig. 3. Time history of angle of tilt

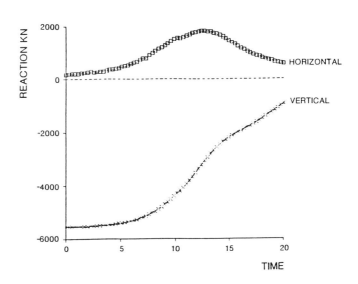

Fig. 4. Reactions at the pivot

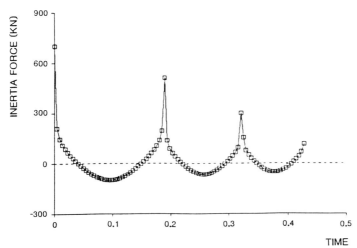

Fig. 5. Inertia force on cylinder

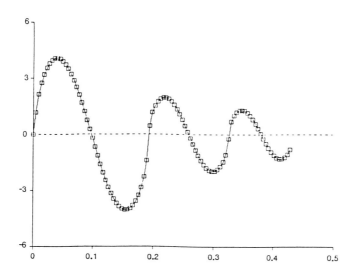

Fig. 6. Drag load on cylinder

NORTH SEA INNOVATIONS AND ECONOMICS

The final set of analyses studied the effects of the explosive loading on members adjacent to the charges. The underwater explosion comprises an initial shock wave, leaving behind a bubble of hot gases; this rapidly expands, but thereafter collapses, re-compressing the gas which then may re-expand, giving rise to a second bubble pulse. This cycle of bubble expansion/contraction may be repeated as many as seven times. Ultimately the internal energy will be dissipated and the gas will disperse.

The structure will be virtually transparent to the initial shock wave, owing to the short duration of the loading event, and the majority of the structural response results from the water particle motions caused by the gas bubble. By mathematical analysis, it is possible to compute the hydrodynamics, and hence to calculate the member loads from Morison's equation. Typical time

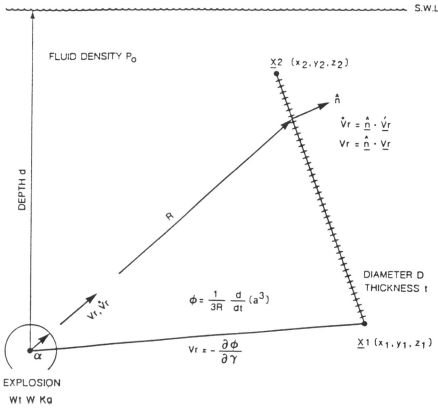

Fig. 7. Explosion loads on cylinder: notation

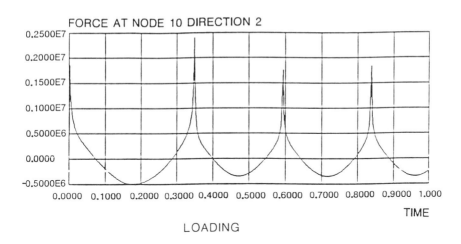

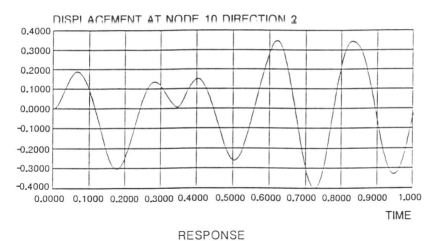

Fig. 8. Response of cylindrical member to bubble pulse loading

characteristics for inertia and drag loading are illustrated in Figures 5 and 6; it may be observed that the majority of the initial loading event has passed within 50 ms of the explosion.

To determine the effects of the loading on the structure, it is sufficient to study the members closest to the charges, since the explosive forces decay rapidly with distance. Each member studied was modelled by a number of beam elements, and the effective loads, which vary both with time and with distance along the member, were applied to the model to perform a full

NORTH SEA INNOVATIONS AND ECONOMICS

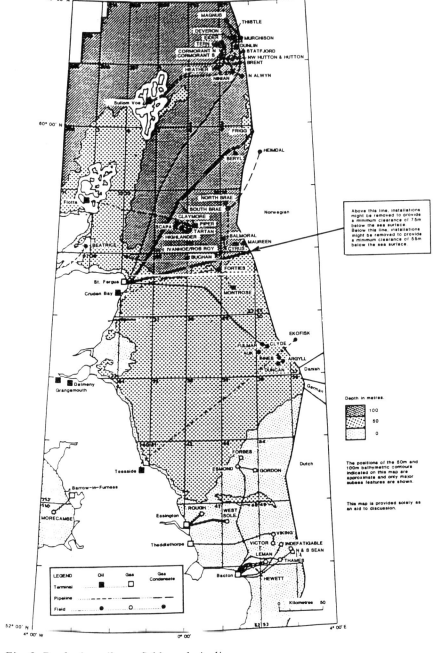

Fig. 9. Producing oil, gas fields and pipelines

transient dynamic analysis (see Figure 7).

Figure 8 shows the loading and response for Brace 'A' of Figure 1. The loading trace shows that the loading follows closely the water particle acceleration as shown in Figure 5, indicating that the inertia force dominates over drag. The peaks of loading correspond to times of minimum bubble radius when the bubble is re-bounding from the collapsed state. These pulses are repeated with decreasing intensity as the energy is dissipated. The bubble pulse period is about 0.35 seconds, which is longer than the natural period of transverse vibration of the member, which is typically about 0.2 seconds. The response of the member is shown on the lower trace of Figure 8. The initial impulse gives rise to a ringing response at the member natural period. The second pulse happens to occur at such a time as to reverse the motion and the third pulse gives the maximum displacement as it is reinforcing the response from the two previous pulses. The maximum displacement for this member at 10m from a charge is about 40cm assuming elastic deformation. It was found necessary to take into account tension effects and the formation of plastic hinges in order to produce a realistic prediction of the member response. Grouting this member reduced the response by about 25%.

A further result obtained from the hydrodynamics study was the pressure transient resulting from the explosion. It was demonstrated that members in the vicinity of the charges would experience peak pressures which are significantly in excess of the hydrostatic pressure. The risk is that the pressure transient could trigger hydrostatic collapse, and it is proposed that critical members should be flooded or grouted to avert this possibility.

Conclusions

In this paper the deconstruction of steel platforms has been examined, with particular reference to platform toppling. It has been shown that by a judicious combination of practical engineering and numerical studies, it is possible to engineer a practicable deconstruction technique. These techniques are currently being used by oil companies who are carrying out studies into platform deconstruction. Toppling is seen to be an attractive solution to the deconstruction and disposal of steel platforms, for sites where it is applicable.

References

1. Final report for joint industry project. Platform abandonment. Wimpey Offshore report WOL 113/88, June 1988.
2. SAUNDERS D.R. et al. Toppling technique applied to platform removal in rigs to reef program. Paper no. OTC 4761, Offshore Technology Conference, Houston 1984.

3. THORNTON W.L. and QUIGEL J.C. Case history for rigs to reefs: a cost effective alternative for platform abandonment. Paper no. OTC 5876, Offshore Technology Conference, Houston 1988.

Discussion

The speaker was asked whether a risk analysis needs to be done in order to ensure that the jacket would topple first time. The speaker replied that it would and one may need to go to a specialist company. However he conceded that SLP did not do one in the study they carried out.

A delegate asked how easy it is to model test the method. The speaker explained that the bubble pulse load is based on potential theory published by Cole in 1949 and that the whole process is predictable, providing the explosion goes off. Care must however be taken to take into account free surface and sea bed effects by the use of 'image' sources and sinks.

Open forum

This forum was chaired by Mr Ron Goodfellow, Mott MacDonald Ltd. The panel members were Mr Trevor Freund of Deminex UK & Gas, Mr Gregory King of Mobil North Sea Limited, Mr John Carter of Shell UK Exploration and Production and Mr Colin Braithwaite of British Gas Plc.

Opening the discussion Mr Freund stated that he felt that the future would see a move to subsea production as there would no longer be the same need for large jackets because of the mature infrastructure in the North Sea. He also felt that there would be a need to develop the more marginal fields, use new technology to extract the oil that has been left behind by conventional methods, and that there should be a move to the partnering type of arrangement so that the above can be achieved.

Mr King stated that his idea of the future would be a change in the current Client/Contractor type relationship. He was concerned as to how new technology would be developed. In the past, joint industry-funded research had failed to benefit the industry.

Mr Braithwaite said that the industry must learn from the Japanese approach to contracting. He added that the industry must learn from their successes and that there would have to be a move towards a more symbiotic relationship between the interested parties in the future.

Mr Carter stated that the future would see a double-thrusted activity in the North Sea. Refurbishment projects would take on more importance with the emphasis switching to the use of existing infrastructure. He added that he saw a change in the existing culture between contractors, consultants designers and owners.

Discussion was then opened to the floor.

A delegate spoke of the need to change exploration wells into production wells, especially for small gas pockets in the Southern North Sea. As this would require a revision of the equipment used he felt that the industry would require ideas different to those put forward by the conference speakers. In response Mr King said that the industry should now be looking to develop reservoirs using the extended drilling techniques, and thus would be looking at a new type of reservoir - namely, one where the long-term information relating to field performance would not be available at the outset.

Another delegate stated that with the drive to reduce CAPEX, it would mean more EPIC contracts and also more partnering arrangements, with the

NORTH SEA INNOVATIONS AND ECONOMICS

very real possibility of partnering for life being an option.

Mr Carter said that there already exist groups which continue to provide the maintenance for 4-5 years after project completion and thus that the industry is already moving towards the longer-term type of contract. He also stressed the importance of accurate specification of scope in invitations to tender and contracts for EPIC contracts.

A delegate was interested in how the cost of a 30 000 tonne crane vessel mentioned earlier could be justified if the industry was moving towards subsea. In response, Mr J. VanBoxtel said that All seas had developed this idea some years back, but the market indicated that it didn't require such vessel and so this idea was shelved. However, a large crane vessel could be used to install or remove large topsides.

One delegate stated that this company had developed a split padear casting which could be used to attach to tubulars for lifts of up to 30 000 tonnes.

The discussion then changed to the topic of platform abandonment with one delegate asking when the operators will start to abandon their platforms and how much they thought the government would allow them to leave behind on the seabed. Mr Carter said that the operators would prefer to leave the platforms there and look at other means of continuing their use. Multi-phase booster pumps maybe needed for subsea developments over 15km from a host platform. However, the future use of some platforms, such as Auk installed in the mid 1970s appeared doubtful.

Mr Freund said that he felt the abandonment of the early 1970s fields was now not far away. In terms of what could be left behind, he said that although current British thinking revolved around the 75m water depth it must be remembered that the European Community had other views on this issue. Mr King said that all indications pointed towards a major platform being abandoned in 1993 or 1994 with the cost of abandonment being met by the government out of back tax paid by the operators. Finally Mr Carter stated that there were of course environmental problems associated with dropping platforms on the seabed and that these needed to be taken into consideration.

Summing up, Mr Goodfellow said that the future would probably mean the use of subsea technology, with refurbishment a big issue. New projects would be smaller with limited budgets making subsea techniques more favourable. He also said that it was important to keep other markets such as the Far East in mind as possibilities do exist there. He then added that costs would be an important issue in the future and that everybody would have to work towards keeping them to a minimum.